CZECHOSLOVAKIA:
THE HERITAGE OF AGES PAST
Essays in Memory of Josef Korbel

Edited by
HANS BRISCH and IVAN VOLGYES

EAST EUROPEAN QUARTERLY, BOULDER
DISTRIBUTED BY COLUMBIA UNIVERSITY PRESS
NEW YORK

1979

EAST EUROPEAN MONOGRAPHS, NO. LI

him as an excellent scholar because of his contributions to the fields of history and political science, and respect was also due to him because he was a wonderful, gentle, and decent human being. Above all, Professor Korbel was one of those rare people who was always ready to help others and who seldom ventured forth into the dark world of criticizing others. His untimely death played a role in the preparation of this volume because his fellow East European specialists shared the editors' desire to print this book as a small tribute to a great man.

The contributions included in this volume are — with the exception of Professor Jan F. Triska's article — all original. Two themes run throughout the articles and the Czechoslovakian heritage which they describe: the heretical tendencies of Czechoslovakians, and the haplessness of Czechoslovakians. The last three articles in this volume (by Svitak, Triska, and Valenta) deal with the Prague Spring, a series of events which illustrates the peculiar mixture of these national traits very well. The appendices contain reprints of some of the heretical political documents which surfaced shortly after the Warsaw Pact intervention and are indicative of the points at which the Czech variant of communism departs from the orthodox doctrine championed by the Soviet Union and its followers.

The contributors, all of whom are practitioners in the field of Slavic and East European studies, need no introduction. Their articles bespeak their attachment to Czechoslovakia, to the Czechoslovakian people, and to Czechoslovak culture.

Hans Brisch and Ivan Volgyes

INTRODUCTION: THE CZECHOSLOVAK TRADITION

The editors, born on the Rhine and on the Danube, have not been involved in Czechoslovakia's struggles, tragedies, triumphs, and disasters; they have been observers — Eastern European specialists whose interest in the fate of Czechoslovakia has been largely academic. Yet this volume came into being because of a special affection for the people of Czechoslovakia, a special sorrow for their circumstances, and a special hope for their future which has developed despite the detachment of the editors. This volume was born because of a conviction that the history and contributions of Czechoslovakia to mankind are great and ought to be known by the outside world.

This volume was also born out of an impatience and frustration that the editors experienced throughout their careers as teachers of Eastern European politics. In trying to make their students understand the area, the editors have frequently had to deal with students whose rudimentary knowledge of the region rarely encompassed more than a single discipline at best, and was then usually confined to a single time period and the works of a single author. To point to other disciplines, other times, other authors and books has meant that the students have had to read at great length about isolated topics and, given the scarce amount of time devoted by students to a single course, there was never a single book available that could be given to them which would adequately cover the area. The desire to fill this gap played a large part in the creation of this volume.

The third impetus for this volume came from an unexpected and unfortunate source, the untimely death of Professor Josef Korbel in July of 1977. His academic stature, his contributions to the field of scholarship and especially his dedicated research which advanced American Slavic studies were well known. The editors had long respected Professor Korbel for his fine teaching which is evidenced by the excellent young professionals he taught. They had long respected

Contents

ACKNOWLEDGMENT

The editors of this volume would like to acknowledge the conscientious and extensive copy editorial work of Mr. John C. Wiltse. His stylistic and textual changes have greatly enhanced the readability of this volume. His careful contribution in checking and rechecking missing or contradictory data made this manuscript a far better product.

The editors of this volume also gratefully acknowledge their debt to Mrs. Jo Ann Leseberg, who typed the manuscripts.

45. *Divide and Conquer: German Efforts to Conclude a Separate Peace, 1914–1918.* By L. L. Farrar, Jr. 1978.
46. *The Prague Slav Congress of 1848.* By Lawrence D. Orton. 1978.
47. *The Nobility and the Making of the Hussite Revolution.* By John M. Klassen. 1978.
48. *The Cultural Limits of Revolutionary Politics: Change and Continuity in Socialist Czechoslovakia.* By David W. Paul. 1979.
49. *On the Border of War and Peace: Polish Intelligence and Diplomacy in 1937–1939 and the Origins of the Ultra Secret.* By Richard A. Woytak. 1979.
50. *Bear and Foxes: The International Relations of the East European States 1965–1969.* By Ronald Haly Linden. 1979.
51. *Czechoslovakia: The Heritage of Ages Past.* Edited by Hans Brisch and Ivan Volgyes. 1979.

20. *Hungary between Wilson and Lenin: The Hungarian Revolution of 1918–1919 and the Big Three.* By Peter Pastor. 1976.
21. *The Crises of France's East-Central European Diplomacy, 1933–1938.* By Anthony J. Komjathy. 1976.
22. *Polish Politics and National Reform, 1775–1788.* By Daniel Stone. 1976.
23. *The Habsburg Empire in World War I.* Robert A. Kann, Bela K. Kiraly, and Paula S. Fichtner, eds. 1977.
24. *The Slovenes and Yugoslavism, 1890–1914.* By Carole Rogel. 1977.
25. *German-Hungarian Relations and the Swabian Problem.* By Thomas Spira. 1977.
26. *The Metamorphosis of a Social Class in Hungary During the Reign of Young Franz Joseph.* By Peter I. Hidas. 1977.
27. *Tax Reform in Eighteenth Century Lombardy.* By Daniel M. Klang. 1977.
28. *Tradition versus Revolution: Russia and the Balkans in 1917.* By Robert H. Johnston. 1977.
29. *Winter into Spring: The Czechoslovak Press and the Reform Movement 1963–1968.* By Frank L. Kaplan. 1977.
30. *The Catholic Church and the Soviet Government, 1939–1949.* By Dennis J. Dunn. 1977.
31. *The Hungarian Labor Service System, 1939–1945.* By Randolph L. Braham. 1977.
32. *Consciousness and History: Nationalist Critics of Greek Society 1897–1914.* By Gerasimos Augustinos. 1977.
33. *Emigration in Polish Social and Political Thought, 1870–1914.* By Benjamin P. Murdzek. 1977.
34. *Serbian Poetry and Milutin Bojic.* By Mihailo Dordevic. 1977.
35. *The Baranya Dispute: Diplomacy in the Vortex of Ideologies, 1918–1921.* By Leslie C. Tihany. 1978.
36. *The United States in Prague, 1945–1948.* By Walter Ullmann. 1978.
37. *Rush to the Alps: The Evolution of Vacationing in Switzerland.* By Paul P. Bernard. 1978.
38. *Transportation in Eastern Europe: Empirical Findings.* By Bogdan Mieczkowski. 1978.
39. *The Polish Underground State: A Guide to the Underground, 1939–1945.* By Stefan Korbonski. 1978.
40. *The Hungarian Revolution of 1956 in Retrospect.* Edited by Bela K. Kiraly and Paul Jonas. 1978.
41. *Boleslaw Limanowski (1835–1935): A Study in Socialism and Nationalism.* By Kazimiera Janina Cottam. 1978.
42. *The Lingering Shadow of Nazism: The Austrian Independent Party Movement Since 1945.* By Max E. Riedlsperger. 1978.
43. *The Catholic Church, Dissent and Nationality in Soviet Lithuania.* By V. Stanley Vardys. 1978.
44. *The Development of Parliamentary Government in Serbia.* By Alex N. Dragnich. 1978.

EAST EUROPEAN MONOGRAPHS

The *East European Monographs* comprise scholarly books on the history and civilization of Eastern Europe. They are published by the *East European Quarterly* in the belief that these studies contribute substantially to the knowledge of the area and serve to stimulate scholarship and research.

1. *Political Ideas and the Enlightenment in the Romanian Principalities, 1750-1831.* By Vlad Georgescu. 1971.
2. *America, Italy and the Birth of Yugoslavia, 1917-1919.* By Dragan R. Zivojinovic. 1972.
3. *Jewish Nobles and Geniuses in Modern Hungary.* By William O. McCagg, Jr. 1972.
4. *Mixail Soloxov in Yugoslavia: Reception and Literary Impact.* By Robert F. Price. 1973.
5. *The Historical and National Thought of Nicolae Iorga.* By William O. Oldson. 1973.
6. *Guide to Polish Libraries and Archives.* By Richard C. Lewanski. 1974.
7. *Vienna Broadcasts to Slovakia, 1938-1939: A Case Study in Subversion.* By Henry Delfiner. 1974.
8. *The 1917 Revolution in Latvia.* By Andrew Ezergailis. 1974.
9. *The Ukraine in the United Nations Organization: A Study in Soviet Foreign Policy. 1944-1950.* By Konstantin Sawczuk. 1975.
10. *The Bosnian Church: A New Interpretation.* By John V. A. Fine, Jr., 1975.
11. *Intellectual and Social Developments in the Habsburg Empire from Maria Theresa to World War I.* Edited by Stanley B. Winters and Joseph Held. 1975.
12. *Ljudevit Gaj and the Illyrian Movement.* By Elinor Murray Despalatovic. 1975.
13. *Tolerance and Movements of Religious Dissent in Eastern Europe.* Edited by Bela K. Kiraly. 1975.
14. *The Parish Republic: Hlinka's Slovak People's Party, 1939-1945.* By Yeshayahu Jelinek. 1976.
15. *The Russian Annexation of Bessarabia, 1774-1828.* By George F. Jewsbury. 1976.
16. *Modern Hungarian Historiography.* By Steven Bela Vardy. 1976.
17. *Values and Community in Multi-National Yugoslavia.* By Gary K. Bertsch. 1976.
18. *The Greek Socialist Movement and the First World War: the Road to Unity.* By George B. Leon. 1976.
19. *The Radical Left in the Hungarian Revolution of 1848.* By Laszlo Deme. 1976.

James B. Bruce

IN MEMORIAM: JOSEF KORBEL

Josef Korbel's untimely death in July, 1977, from a vicious form of pancreatic cancer, cut short the career of an inspirational teacher-scholar at his best, and the life of a kind and gentle man, admired and loved by those of us lucky enough to have known him. His many and varied accomplishments bespeak a life of singular determination, rare ability, and uncommon humaneness. These accomplishments alone would make for a distinguished career. But Josef Korbel had two careers: he was an extraordinary diplomat and a great teacher-scholar. In both careers, his considerable talents and efforts brought him much deserved recognition.

In retrospect, it seems almost inevitable that Korbel the academic presupposed Korbel the diplomat. He probably did not plan it that way — and certainly not in the way it happened — but the quality of his scholarship leaves little doubt concerning the aphorism that wisdom presumes both knowledge and experience. Josef Korbel had both, in ample measure, and suitably combined. It was this happy synthesis, and a good many other qualities as well, that mark his many contributions as uniquely worthwhile.

He demonstrated his abilities early. Born not far from Prague in 1909, Josef Korbel did well in school. He spent his twentieth year at the Sorbonne, and, returning to Prague, received his J.D. from Central Europe's oldest and most pretigious institution, Charles University, in 1933. As a student of international law and a journalist during his college years, Korbel's interest in the uncertainties of the European political world were more than academic. Following a year's stint as a first lieutenant in the Czechoslovak Army, he achieved his long-desired ambition of entering the Diplomatic Service. That he did so initially at no pay, and after declining a lucrative opportunity with a private firm, reveals much of the selfless dedication that marked his many professional endeavors in the years that followed.

Thus Korbel began, rather inauspiciously at that, a first career. In the ensuing fourteen years, Korbel's diplomatic talents would take him from Prague to Belgrade, London, and Paris, as well as New Delhi and Karachi. It was in such places, and in service to humane and democratic ideals, that some of the major themes of his later teaching and writings were forged.

Korbel's diplomatic assignments were not merely responsible ones, they suggested a future of considerable promise. Within five years of joining the diplomatic corps, he was appointed personal secretary to Foreign Minister Jan Masaryk. These were turbulent years. Following the Munich "maelstrom," as he later referred to it, and the dismemberment of Czechoslovakia, Korbel headed the Broadcasting Department for the London-based Government-in-Exile during the war. Returning on the first plane to liberated Prague in 1945, his first assignment as Masaryk's *Chef de Cabinet,* was to help organize and rebuild the Foreign Ministry. Following this, as Czechoslovakia's delegate, Korbel served as President of the Economic Commission for the Balkans and Finland at the Paris Peace Conference in 1946. But his principal responsibility in the post-war period was to represent the Beneš government in Belgrade, where he had earlier served as press attaché in 1936–37. At this time the Yugoslav post was a sensitive and critical assignment. And it was also something of a milestone: at 36, Josef Korbel became the youngest ambassador to serve the Czechoslovak Republic.

It was Korbel's direct experiences in the cascade of events of the late 1940s that provided the focus of his early writings. These concerned Yugoslavia, Czechoslovakia, and Kashmir, all arenas of major political consequence. As a participant in the international politics of the period, his involvement thus entailed both a European phase, as Czechoslovak ambassador, and a later Asian phase, as UN Representative. As a European statesman, Korbel ably represented Czechoslovak interests in Belgrade during the fateful years of the establishment of Communist governments throughout Eastern Europe. From this experience came *Tito's Communism* (University of Denver Press, 1951), a penetrating and critical portrait of a new and repressive regime. Not only was the volume one of the earliest on the subject, it remains valuable as a first-hand account of one who had direct access to the major political figures. But beyond the careful observations recorded there, Korbel's tenure in Yugoslavia provided invaluable experience for the assignment that followed.

Shortly before the Communist seizure of power in Prague in February 1948, Ambassador Korbel was named to the United Nations Commission on India and Pakistan (UNCIP), the Security Council's first efforts to mediate the explosive Kashmir crisis. Owing to the intense political and religious passions of those involved in the hostilities, quick solutions were simply precluded. As Chairman of the Commission, Korbel not only won the confidence of its members (throughout the 113 meetings held during that period, every decision was unanimous), but of the antagonists as well. In the face of repeated frustrations and setbacks, the Commission finally succeeded in persuading the Indians and Pakistanis to accept the outcome of a plebiscite in Kashmir, and, effective on July 1, 1949, to accept a cease-fire after fourteen months of bitter armed hostilities. For such an intractable problem, the event was significant. In Michael Brecher's study of the crisis, he judged that cease-fire as "the most striking UN achievement."

Korbel's replacement on UNCIP by another Czechoslovak delegate, representing the new Communist government, marked the end of his diplomatic career. And to the good fortune of the world of scholarship, he promptly began another. Now an exile, Josef Korbel found himself in New York with Mandula, his wife for fourteen years, and their three children. Through the good offices of Columbia's noted Soviet specialist Philip E. Mosely, Dr. Korbel secured in 1949 a temporary one-year appointment with the Social Science Foundation and the University of Denver's Department of International Relations. Here too he successfully combined his enormous energies and creative skills to the lasting benefit of all who associated with him.

Josef Korbel thus began his second career at the age of forty, an exile in an alien land, virtually broke, and a neophyte in the mores of American academe. Yet in every phase of his new profession — in teaching, in research, and in administration — he excelled.

His first three books integrate with skill and sensitivity the scholar's quest for understanding with the insights afforded only to participants and first-hand observers. After *Tito's Communism,* Korbel's next volume was *Danger in Kashmir* (Princeton, N.J.: Princeton University Press, 1954; rev. ed., 1966). It stands today not only as the best account of the UN efforts to mediate the hostilities, but also as a careful and balanced study of the origins of the crisis and the issues impeding its resolution. His third and final work in this genre is *The Communist Subversion of Czechoslovakia, 1938-1948: The Failure of Coexistence* (Princeton, N.J.: Princeton University Press, 1959). Here, with the perspective of a decade of reflection, and the vantage point of his close access to many of the political figures (including Beneš) involved in the weeks just prior to the coup, Korbel reconstructs the tragic throttling of Czechoslovak democracy. What emerges is an absorbing mosaic of Communist duplicity and resolve, indecisive and divided government leadership, and the oppressive weight of the Munich legacy. Some nagging questions and moral dilemmas posed here anticipate his powerful *Twentieth Century Czechoslovakia: The Meanings of Its History* (New York: Columbia University Press, 1976), published shortly before his death. But between his two significant contributions to our understanding of Czechoslovakia is an impressive catalog of other contributions, some in writing, others in teaching and administration.

From 1959 to 1969, Josef Korbel set his energies to building a solid graduate program in international relations at the University of Denver, and the only program of its kind in the vast region between the Mississippi River and the west coast. As Director of the Social Science Foundation, and later, the first Dean of the new Graduate School of International Studies, Korbel became, as his successor Robert C. Good has aptly noted, an "indefatigable builder." With consummate skill, Korbel raised an astonishing three million dollars during this period, recruited a talented faculty, and shepherded a promising idea into a unique and exciting program of international studies.

His teaching and research continued, of course, and in 1963 appeared his most accomplished diplomatic history: *Poland Between East and West: Soviet and German Diplomacy Toward Poland, 1919–1933* (Princeton, N.J.: Princeton University Press, 1963). In this work, Korbel examines the most enduring geopolitical determinant of the shape of modern Europe. A decade later, he returned to the complex German-Russian equation, this time in the larger context of a contemporary and changing Europe. In *Detente in Europe: Real or Imaginary?* (Princeton, N.J.: Princeton University Press, 1972), the first major study of this controversial issue, Korbel cautioned that the removal of mistrust — requisite to a durable detente — presumes a measure of liberalization in the Communist states. That he perceives this as a genuine possibility is itself worthy of note. But it also reveals something of the strength of his belief that man's hunger for freedom and individual dignity will somehow prevail. As he phrased it, this "perennial struggle . . . is both man's uniqueness as man and the ultimate justification for his existence."

Numerous awards bestowed on Josef Korbel have recognized his many scholarly accomplishments. Among those connected with his research and writing include various appointments and fellowships at Harvard, M.I.T., Columbia, and Oxford; support from the Rockefeller and Guggenheim Foundations; and other grants from the Social Science Research Council, the American Council of Learned Societies, and the National Endowment for the Humanities. To these research awards can be added the recognition he justly received for outstanding teaching, including University of Denver Lecturer in 1958, and Outstanding Educator of America in 1971. On retiring from the Deanship in 1969, he was awarded the chair Andrew W. Mellon Professor of International Relations, and in 1975 was suitably honored by his university with the award of an honorary Doctor of Humanities, and with the establishment of a graduate fellowship in his name.

As a teacher, Josef Korbel was superb. To have attended his lectures on the political consequences of World War I in Europe, or on the origins and meaning of the Prague Spring, to cite but two memorable examples, was an intellectual experience of the highest reward. Yet for all his direct encounters with the political events attending his first career — indeed, for one who had fairly extensive contact with such figures as Jan Masaryk and Beneš, and at times Tito and Nehru — he was never anecdotal, nor trivial. His command of detail was awesome, but of greater moment was his genius for moral insight. Only rarely does one encounter a truly gifted teacher, and there was never a doubt among those of us who knew him at Denver. As Charlotte Read, one of his most recent students, expressed it, Professsor Korbel was "always hoping to challenge those who felt themselves infallible, and to encourage those who felt themselves inadequate. . . . It seems his purpose in teaching was to entice the inquisitive mind and to develop [his] craving and incredible perseverance in his students."

Indeed, as a teacher-scholar, Josef Korbel synthesized the best of traits: with a passion for learning and devotion to truth, he combined meticulous research and gifted teaching. But judiciously wedded to rigor and discipline was an unyielding humanism, as evident in his personal concern for students as in his interpretations of the complex events examined in the six books and many journal articles he authored.

Above all else, Josef Korbel loved freedom, and the ennobling dignity of the individual that it alone can nourish. Nowhere in his works is this more clear than in his two volumes on Czechoslovakia. Yet neither is polemic or pious. Korbel's searching interpretation in *Twentieth Century Czechoslovakia* confronts, inescapably, the enduring paradox of Czechoslovak history: why has a nation of such notable democratic and social achievements seemed to cooperate in its own demise, and with such apparent willingness? The question is as painful as the answers are complex. But the critical events of Czechoslovak statehood — those of 1938, and 1948, and again in 1968 — all suggest a disturbingly similar pattern: the political leaders failed "to perceive any alternatives to submission." We are thus left to ponder: Is freedom, if easily gained, easily lost?

With remarkable objectivity, Korbel examines the contending choices then available, and concludes that the alternatives to repeated capitulation were indeed viable. It is a compelling case. But it is not speculative history. For the basic thrust of the logic — and in this respect, it is so much like Korbel himself — is to look forward, not backward. And *Twentieth Century Czechoslovakia,* like *Detente* before it, ends on a sanguine note: the human urge to freedom is too strong, too resilient, to be suppressed forever. It is a great — and plausible — expectation.

But doubts, too, seem warranted. Perhaps the most convincing of the volume's principal theses is that of the persisting inadequacy of the Czechoslovak political leadership at moments of threatened peril. With the conspicuous exception of Thomas Garrigue Masaryk, who commanded Korbel's highest admiration, these were men, whatever their other virtues, who were not equal to the enormity of the crises they faced. In all cases, political freedom was the first casualty.

Reflecting for a moment on the distinguished accomplishments of Josef Korbel, the American student of the Czechoslovak dilemma cannot escape the question that the emigré cannot easily raise: if the human costs of oppression in Czechoslovakia were somehow calculable, what figure must be assessed for the loss of Korbel? This *is* "what if" history: In terms of the diplomatic and political leadership he might have afforded, or the lives there he might have enriched, what invaluable contributions have been denied a small nation that surely could have profited from them?

Put this way, one of the meanings of Czechoslovak history has to be that our gain was Prague's loss: the inestimable value of Josef Korbel. What price oppression? In a way that the *jus soli* American can never truly understand, Josef Korbel did not take his freedom for granted. And for this we should be profoundly grateful, and perhaps occasionally reflect for a moment on what it means to us.

BIBLIOGRAPHY OF JOSEF KORBEL'S MAJOR WRITINGS

BOOKS

Twentieth Century Czechoslovakia: The Meanings of Its History. New York: Columbia University Press, 1977; 346 pp.

Detente in Europe: Real or Imaginary? Princeton: Princeton University Press, 1972; 302 pp.

Poland Between East and West: Soviet and German Diplomacy Toward Poland, 1919-1933. Princeton: Princeton University Press, 1963; 321 pp. Also in paperback edition.

The Communist Subversion of Czechoslovakia, 1938-1948: The Failure of Coexistence. Princeton: Princeton University Press, 1959; 258 pp. Also in paperback edition, and in Japanese and Spanish translations.

Danger in Kashmir. Princeton: Princeton University Press, 1954; 351 pp. Revised edition published in 1966.

Tito's Communism. Denver: University of Denver Press, 1951; 368 pp. Out of print.

ARTICLES

"Detente and World Order," *Denver Journal of International Law and Policy,* VI (Spring, 1976), pp. 9-18.

"West Germany's Ostpolitik: II, A Policy Toward the Soviet Allies," *Orbis,* XIV (Summer, 1970), pp. 326-348.

"West Germany's Ostpolitik: I, Intra-German Relations," *Orbis,* XIII (Winter, 1970), pp. 1050-1072.

"German-Soviet Relations: The Past and Prospects," *Orbis,* X (Winter, 1967), pp. 1046-1060.

"Changes in Eastern Europe and New Opportunities for American Policy," *World Politics,* XVIII (July, 1966), pp. 749-759.

"Czechoslovakia," in Thomas T. Hammond, ed., *Soviet Foreign Relations and World Communism,* Princeton: Princeton University Press, 1965, pp. 312-333; co-editor of annotated bibliography.

"The Kashmir Dispute," in A. Gyorgy and H. Gibbs, eds., *Problems in International Relations,* Englewood Cliffs, N.J.: Prentice-Hall, 1955, pp. 160–168.

"The National Conference Administration of Kashmir, 1949-1954," *The Middle East Journal,* VIII (Summer, 1954), pp. 283-294.

"Danger in Kashmir," *Foreign Affairs,* XXXII (April, 1954), pp. 482-90.

"The Kashmir Dispute After Six Years," *International Organization,* VII (November, 1953), pp. 498-510.

"Some Aspects of the Internal Development in the State of Jammu and Kashmir," M.I.T. Center for International Studies, Summer, 1953; 56 pp.

"Titoism: An Evaluation," *Journal of Central European Affairs,* XI (January–April, 1950), pp. 1-9.

"The Kashmir Dispute and the United Nations," *International Organization,* III (May, 1949), pp. 278-287.

In addition, shorter articles and essays by Josef Korbel have appeared in such publications as *East Europe, Foreign Policy Bulletin, Far Eastern Survey, The New York Times, New Leader, Worldview, Commonweal, Behind the Headlines,* and *Vital Speeches.* He presented dozens of research papers to meetings of professional societies, and delivered a like number of major addresses to varied scholarly gatherings at many colleges and universities. At the time of his death, Professor Korbel had all but completed research for his seventh book, tentatively titled, *A Trans-Siberian Anabasis: The Czechoslovak Legionaires.*

Michael G. Fry

JOSEF KORBEL: HISTORIAN

Politicians and officials prosper but rarely in academic life; academics, by and large, make ineffective practitioners. The costs of entry and re-entry are invariably high. Transplanted men of government prefer memoranda to books, the press to theses, and reminiscences to analysis. Uprooted professors, even when they have uprooted themselves, seem unnecessarily theoretical, excessively esoteric, and un-realistic to the point of naivete. The former seem hard-bitten and all too rigid; the latter, crusaders who bruise too easily.

Yet it is not always so and Josef Korbel was one of those who crossed the boundaries with ease en route to a long, secure, and successful career. In 1949, at the age of forty, an emigré from his native Czecho-slovakia and a refugee from communism, he became a professor of international studies at the University of Denver. He was by training a lawyer, by vocation a diplomat, and by adoption a scholar. Between 1951 and his death in July 1977, he published six books, five of which dealt with twentieth century Eastern Europe and are discussed here in terms of their unifying themes.[1] But first the man and his art as re-vealed in his writing.

Korbel's legal education left distinct and particularly relevant marks. He was deeply concerned that political leaders uphold the law. He demonstrated an unfailing devotion to legal processes, and under-stood the profound relationship that must exist between law and justice. An independent judiciary, maintaining the rule of law, was, in his view, an indispensable part of a democratic society. He expected, therefore, the Supreme Court of Czechoslovakia to resist Communist policies, as it did in February 1948 over the issue of nationalization; and that Prokop Drtina, as Minister of Justice, would oppose Communist infiltration and subversion of the judiciary. Violations of the law and disdain for legality were for him fundamental tests. The "revolutionary justice" of postwar Eastern Europe was, therefore, utterly alien to him.[2] In their persistent violations of the law he found perhaps the most telling basis from which to judge and condemn the Communist governments of Eastern Europe.

Furthermore, the conceptual framework from which, particularly, he wrote international history came from the law. International history was for him traditional diplomatic history, *l'histoire des traités,* with the outputs of policy, treaties, formal agreements, and understandings, at the core. Treaties were the landmarks of history. History unfolded as they were signed, then honored or broken. What morality or immorality existed in international affairs, what stability or instability persisted in the international system, stemmed from that process.

Korbel the diplomat surfaced in several ways; a certain elitism, a concern for propriety and correct behavior, and a skeptical distance from mere politicians. Czechoslovakia's first Communist Prime Minister, Klement Gottwald, for instance, was not a gentleman. He was crude, vulgar, socially inferior and, indeed, can a Communist, disloyal, dishonorable, unreliable, without either scruple or dignity, be a gentleman? Quiet and careful diplomacy, and the proper use of experts in, for example, the preparation of conferences, were infinitely more valuable than oratory, public diplomacy, and the flamboyance of state visits. In essence, this was a plea for order and professionalism in the conduct of international affairs that one would expect from a professional diplomat; a plea for pragmatism and realism in the derivation of policy.[3] It followed that Korbel would treat the likes of General de Gaulle with unmistakable condescension. The intellectual links that frequently exist between the diplomat and the elite journalist are also in evidence; witness Korbel's eye for detail, his powers of observation, his vivid memory, and his sensitivity to accurate description. Equally, he knew when politicians used and abused the media and denied it freedom; when newspapermen ceased to be journalists, when publishers served only the state.

Korbel the scholar ranged from memoir through traditional political, and diplomatic to contemporary history. He wrote in an essentially narrative and descriptive fashion, but challenged readers with his forthright interpretations. There was, therefore, a forceful and telling simplicity about his work. He never approached a subject historiographically and rarely debated evidence, and yet he demonstrated in one particular way a singular sense of the appropriate. That is to say, his memoirs and contemporary history were not overburdened by an unnecessarily ponderous scholarly apparatus; footnotes were not piled on each other like firewood for the winter. Yet his political and diplomatic histories were based solidly on research, an awareness of the limitations of his sources, and a determination not to place more weight on the evidence than it could bear. These properties were demonstrated for the first time in a complete way in his volume on Poland and at their fullest in his final book, *Twentieth Century Czechoslovakia.*

Some qualities never changed; others grew and altered, and were reinterpreted. Korbel the emigré, the advocate, lacing his work with autobiographical content, inserting a personal treatise, judgment, testimony, or warning, was always present. He was able, therefore, to arrive with ease at that eminently sensible view that a work of history can be neither definitive nor objective. History and personal experience were for him intimately interwoven for ". . . a scholar inescapably reads the historical record in much the same way as he would look at a mirror — what is most clear to him is the image of his own values, his sense of his own identity."[4] He wrote, therefore, consciously and deliberately, "'neither without love nor without anger' for, as Karl Marx put it, ' truth is not without passion, and passion is not without truth.'"[5] His superb command of West European and Slavic languages was always in evidence, giving his work both authority and sophistication. Finally, and with feeling, he remained convinced of the intellectual frailty of certain endeavors in the social sciences, and of the ultimate futility of seeking explanation in the scientific sense, derived from quantitative analysis, for the richness of human experience.

> No computer will ever be able to measure the enormity and the depth of suffering of the people of the East at the hands of Hitler's Germany.[6]

and

> There is no scientific way to trace or test either the precise nature or the exact degree of the influence on a people of their history.[7]

Equally, he was no worshipper at the altar of definitional rigor and conceptualization, preferring to believe, in the European tradition, that concepts constrain more than they liberate.

Growth and change are, however, equally discernible. From an occasional witty or striking phrase in his earlier works, via a certain dramatic quality, for example in the volume on Poland, he developed a prose style that was usually clear and sometimes moving. His last work, *Twentieth Century Czechoslovakia,* has an emotional quality of some power with metaphors and similes flowing from his love of natural and physical beauty. Clearly, he was convinced that biography was not history and initially, he was more at ease with groups, elites, political parties, and collectivities acting as social and political forces than individuals. Marshal Tito, for example, was presented as the most profound expression of the Partisan psychology.[8] There were relatively few striking portraits and fewer profound explorations of the layers of the individual personality. In the book on Poland, however, high politics come to life. Individual statesmen and their records, memoranda, notes, *procès-verbaux,* assume a central role in the unfolding of history. In his final work, individuals and groups are in mature balance. Initially, also, his use of political data greatly outweighed, for example, economic data, even though he ridiculed Tito's first Five Year Plan unmercifully. In his work on detente, however, the research base broadened. He used economic, socio-cultural, and technological data to a far greater and more telling degree, demonstrating a broadening of range and capability.

The larger question remains. What was the point of historical scholarship? Korbel seemed to harbor few doubts. First, the historian must identify and explain patterns in history, patterns of human behavior, experience, and response that were replicable. Here lay the utility of history, illuminating the past, bringing order, and even endowing it with predictive qualities. Uppermost in his mind, perhaps, was the pattern of Communist subversion of democracy in Eastern Europe: assault from within using familiar processes, institutions, language, and traditions to reassure the prey; infiltration of police, army, bureaucracy, media, and government; creation of national organizations at all levels; seduction of the intellectuals; development of an illegal military force; ensuring the cooperation of Russia; relying on that eminently reliable factor, the absenteeism of the West; and finally the coup, "a combination of psychological terror, physical violence, and vulgar denunciation of opponents."[9] What might be called a structural pattern complemented this process of Communist subversion, and, indeed, set the conditions of its success. Any state experiencing economic misery and social injustice, lacking a deeprooted democratic spirit, and showing a willingness to accept dictatorial rule invites the coup de grâce of Soviet military intervention. The exception to this structural pattern in Eastern Europe, and the last state to succumb, was Czechoslovakia, a fact which was an undoubted source of pride to Josef Korbel. Conspirators and those who failed to oppose them cut down his homeland, a theme to which I shall return.

For Czechoslovakia, with the events and decisions of 1938, 1948, and 1968 requiring explanation, Korbel saw a complementary and equally compelling pattern; the pursuit of humanistic ideals and social justice alternating with the willingness to sacrifice both just to survive, even in subjugation and servility. The pattern was to be explained in terms of roots, identities, and the cumulation of Czechoslovakia's historical experience.

> There can be little doubt that the almost 300 years of the Hussite and pre-Hussite tradition, which is generally viewed as a period of national glory, cannot be erased from the minds of subsequent generations. . . .
>
> By the same token, however, 300 years of subjugation, followed by another 100 years of precarious struggle to reawaken the supine nation and its dormant spirit, cannot but infect the fountain of inspiration and derange the compass needle. . . .
>
> As a result, a pattern seems to have developed: in times of assured statehood the Czech people strove for the ideals of humanity; in times of peril, they lowered their heads to save the national body.[10]

From patterns one issued warnings; that trust and tolerance meet violence and treachery in an unequal struggle and inevitably succumb; that faith and hope will not buy peace and freedom; that decency and goodwill cannot match terror; that the West must devise new ways to counter communism or risk collapse; that coexistence will not ensure security; that detente is, at best, a volatile commodity at least as dangerous as it is a source of optimism. Here was Korbel, the child of Munich, perpetually fearful that the West would weaken, appear less vigilant, relax and become both complacent and divided, and pursue that greatest of all illusions, Communist good behavior, for "Do good to the devil, he will repay with damnation."[11]

Finally, the historian retraces the tragedies, more Greek than Christian, of Eastern Europe, Czechoslovakia, and of Eduard Beneš, tragedies from which there seemed little relief and then only in statements of paradox. Eastern Europe, despite its sacrifices in World War II, was denied its freedom. The majority of the people, the agricultural, professional, artistic, technical, administrative, and business classes and groups defined salvation as a break with Russia's abuses and exploitation and as a return to self-administered political pluralism, but they were ignored. Democratic Czechoslovakia, resting uneasily between Germany and Russia, was the victim of a multifaceted conspiracy: the intent and design of the Czech Communists;

the egregious errors, the supreme folly, and the unmistakable treachery of the West on at least three occasions from 1938; [12] Beneš's own policies, and naivete, placing faith in Neville Chamberlain, Stalin, and Gottwald, betrayed by his own trust, loyalty, and devotion to principles; [13] Nazi German and Soviet aims enforced by the *Reichswehr* and the Red Army; and the weakness, disunity, ineptness, and idealism of Czechoslovakia's democratic forces, lacking adequate discipline, plan, leadership, and prescience. The well-meaning Beneš faced a series of extraordinarily difficult situations which offered only costly alternatives as solutions, such as communist rule or civil war in 1948. Acting constitutionally and decently, manacled by his convictions, and tragically error prone, he succumbed to pressure in September 1938, December 1943, March 1945, and finally in February 1948.

This was the Czechoslovakia and the Beneš of Korbel's *The Communist Subversion of Czechoslovakia* written in the 1950s. In his last work, *Twentieth Century Czechoslovakia,* he returned to judge Beneš, more in sorrow than anger, and to compare him with the elder Masaryk. On the occasions when Beneš wavered and faltered, Masaryk would have stood and fought. Beneš had become, over the years of Korbel's study, a complex and even devious figure, weak but not naive, a capitulator whose methods tainted his goals, and whose skills undermined his dignity and principles. As Korbel wrote, "Beneš placed his reliance on his allies and on his skills at diplomacy; Masaryk placed his on what he regarded as historic Czechoslovak moral principles." [14] Here was a different if lesser tragedy, a professed Benešite who could no longer serve Beneš's memory.

Josef Korbel was more of a nineteenth century liberal than a twentieth century democrat. His socio-political philosophy, as it applied to the functioning of the nation state, was that of an educated, successful man with endowed and earned superiority who, for example, had little faith in the political acumen of the industrial worker and rather more sympathy for the common sense of the peasant farmer. As a member of Beneš's eminently respectable National Socialist party in the inter-war years, he felt himself to be a leftist and a liberal progressive, but he held back from radicalism and was repelled by both Communism and Fascism. He regarded himself as a humanist and pluralist, believing in the eminent morality and common sense of moderation. The goal of political life, the raison d'être of the modern state, was to achieve social justice in the true sense which was ". . . the opportunity for each citizen to find the satisfaction of self-fulfillment, to make choices among real alternatives, to develop voluntarily his social consciousness, to make willingly an ethical commitment to his society and to himself, and to discover and be able to act in a personally rewarding life style."[15]

His was an uncomplicated political typology; the Center confronted the extremism of both the Left and the Right. He was unyielding in his condemnation of both Fascism and Communism and morally saw no difference between them. Fascists, to cite Thomas Masaryk, were "pathological scum," a singularly preemptive phrase. Predictably, however, the theory and practice of Communism received fuller attention. Communism was to Korbel many things, and none of them deserved praise. His analysis and judgment of the Marxist-Leninist legacy provides the first dominant and powerful theme of his work.

Communists were the enemies of freedom, democracy, justice, privacy, individual values, and human dignity. They were ill at ease with intellectual honesty, cultural achievement, and political choice. They were guilty of a perpetual distortion of the truth and of a repeated rape of history, and their problem was the singularly difficult one of attempting to predict the past.[16] Communists were purveyors of myth, fallacy, distortions and idolatry of symbols and propaganda. They practiced delusion and self-delusion, largely by way of "a megalomania of figures."[17] In some ways he believed, like Raymond Aron, that Communists were utterly flexible and even brilliant opportunists in terms of means and absolutely dedicated in terms of goals. They were driven by ideology and a new ethic couched in a novel vocabulary of politics and law, which brought them to the point of amorality. On the other hand, and a theme which bore much repetition, they had proved to be both feeble and corrupt in their war-time resistance to the Nazis.

The Communist Party, in or out of power, represented order enforced through fear and discipline imposed through abuse. It luxuriated in a mindless uniformity, a deathly regimentation, a suffocating control, and a ruthless censorship. Pathological distrust marked its members, each of the other, and resulted inevitably in intolerance, arbitrariness, intrigue, exposure, denunciation, persecution, vengeance, arrest, and death. Lies, treachery, betrayal, hypocrisy, and cynicism symbolized the politics of vulgarianism, practiced by vulgarians and their docile sycophants. Yet, of course, in comparison with their democratic enemies, there was organization, skill, plan, cohesion, discipline, leadership, and unity, which led so often to victory.

The Communist State was marked by corruption, drabness, estrangement, alienation, fear, suspicion, fanaticism, tyranny, and terror. It was a totalitarian state, floundering in bureaucracy and regimentation, proving itself to be inefficient to the point of incompetence, wasteful and clumsy to the point of absurdity. This is why, in Korbel's view, the Communists knew they would not triumph if free elections were held in Czechoslovakia in 1948. He had, as I have said, derided Tito's Five Year Plan as theory rich and achievement poor, but even when Communist states made real and perceptible economic advances they could not impress Josef Korbel. Their economic achievements were simply not worth the cost imposed on the individual and society. Predictably, he denounced the principal Communist state, the Soviet Union, because it demanded subservience and understood only dependency relationships in its dealings with other states. This led, in Korbel's view, to a form of moral insanity which persisted in Eastern Europe and yet followed logically from the acts of conspiracy, subversion, and armed intervention.

In the final analysis, communism was the negation of the values of western civilization, an alien, unnecessary, and unacceptable force for evil. It was a disease which attacked the vitality of the democratic system; its corrosive properties ate away at the social, economic, and political fiber.

> The communism that was to have freed man from the fetters of capitalism, shackled him instead with the multiple chains of personal frustration, cultural starvation, economic insecurity, and social repression.[18]

A second grand theme is the struggle in Korbel's mind, over time and across issue areas, between a fundamental optimism and a pessimism which bordered at times on fatalism. This theme is linked directly with his view of communism and his search for both patterns in history and his analysis of tragedy.

Communism, forged in war and entrenched by victory, was indestructible unless recast in the course of another great war. Communist regimes, enjoying all the apparatus of control, the police, the army, the fruits of technology, and a monopoly of both information and political power, easily subdued peoples who were tired, worn down, and without hope. Dissent, therefore, even when presented, for example, with the evidence and the opportunity of economic failure, could not succeed in peacetime. This was Korbel's gloom of the early 1950s, and was not a transient property, for one must question whether he felt that democracy could defeat communism in the circumstances of the mid-century. Yet, even then, he declared that dictatorships, inept in assessing the psychology of a society, could never feel safe, and could retain control only through the escalation of controls, and maintain order only by increased repression.[19] Eventually, he seemed to say, political turbulence would develop and reach the force of revolt. If such regimes resorted to war in order to deflect this threat, then dissent might triumph.

Pessimism marked his analysis of the prospects for detente in Europe in the 1960s. In Korbel's opinion, when reality triumphed over illusion, people would realize that the accommodation between the two political systems which divided Europe would be at best limited and of doubtful duration and at worst simply dangerous. The Soviet invasion of Czechoslovakia in August 1968 was reality; the ideological chasm and the conflicting economic and social systems were reality; Soviet determination to use detente to lull and trick the West was reality. The dangers of detente, therefore, were unmistakable. In a limited way and in the short run, however, certain valuable results could stem from detente such as increased trade, technological cooperation, and cultural enrichment, but even there the Soviet bloc gained more than the West. Korbel did not, therefore, place great faith in theories of convergence and he could not have ended his work on detente on a more sombre note. Only if the East were liberalized, only if men sought their freedom, as they surely would, only if Communists ceased to be Communists, would relationships of confidence and trust be created.

Czechoslovakia's history, in Josef Korbel's opinion, could best be understood in terms of paradox, of a fundamental struggle between conflicting forces and contradictory behavior. He sought to interpret Czechoslovakia's history in terms of the roots of the nation's inheritance, the soil of those roots, the ideas leading to its existence, the loss or contamination of those ideas, the consciousness of the nation's accumulated experience, of its identity transmitted across generations, its characteristics, will, values, and disposition; in sum, in terms of the *forces profondes,* the meanings, of Czech history. On the one hand, and the source of pessimism, stood the record of abdication. The Czechoslovaks, petty, malleable, intriguing, servile, and corrupt, so often in their history experienced a loss of nerve and will, capitulated and agreed to live with oppression and without freedom, renouncing or failing to defend their fundamental values. Czechoslovaks had enjoyed, therefore, only brief interludes of freedom between long periods of Hapsburg, German, and Soviet domination, and thus progress had been followed by dismal retrogression. As a result, they had never retained a national self-confidence, never been fully "deAustrianized" and never developed a permanent dignity, a deep confidence, and a spine of assurance.[20] On the other hand, and the wellspring of both pride and optimism, there stood the record of achievement. The Czechoslovaks, in a determined quest for social justice, cultural growth, personal freedom, humanistic ideals, and liberty, demonstrated the truest instincts, and gave Josef Korbel grounds for idealism and faith. Indeed, in rare and not necessarily weak moments, he would speak of their uniqueness and superiority in comparison with the other nations of Eastern Europe.

As he weighed these forces and used them to explain the crises of recent Czech history, Korbel explored complementary issues. "Does a nation have a moral obligation to defend its rightful position against violence even in the most adverse circumstances? Or is it morally justified in attempting to assure its biological survival . . . at the cost of even the temporary loss of its moral integrity and fundamental values?" His answer, for example on Munich, ". . . the valiant ethos of the nation demanded from its leaders the ethical, not the practicable position. The Munich dictate should have been rejected, no matter what the consequences."[21] Will the freedom won in combat have greater vitality than that secured through negotiation? Why is suffering so often the finest source of regeneration; why does hope spring from despair?

At one point the central theme is laid out starkly and pessimism seems to triumph.

> There remains in the national tradition of Czechoslovaks the kind of ethnic memory of a former idealism and heroism which, like seeds, lie dormant in the soil, but which, given a bit of the sun of tranquillity, flowers into a passion for freedom and social justice unusual in kind and degree in all of Eastern Europe; there also remains a conditioned reflex for survival, nurtured carefully during 300 years of oppression, in which every trick of cautious advance and hasty retreat has been carefully explored and well learned; when this passion for freedom and social justice, and the will to survive are suddenly opposed in separate and conflicting orbits of events, survival becomes the first law.[22]

Josef Korbel, like Ralf Dahrendorf, never felt that mere survival was enough; man must go forward in dignity and freedom.

Finally and firmly, however, and almost in defiance of much of his own evidence, Korbel restated and reaffirmed his belief in the finer qualities of his fellow countrymen and of mankind. The authentic moral and spiritual health, and the fundamental democratic instincts of the Czechoslovaks would, in the long run, triumph. A Czechoslovak renaissance was probable. The spring of 1968, Alexander Dubček's spring in search of socialism with a human face would flower again, the lindens of freedom would burst forth as the minds of men reasserted themselves over an alien system, as the nation rediscovered its spirit.

> The spark is still there. One cannot doubt that it will flicker one day again into flame, and freedom will return to this land that is so essentially humane. And if again it dies, unprotected by the leaders and undefended by the people, then it will arise again and still again.[23]

A third, if muted theme, for Korbel wrote only one diplomatic history,[24] concerns the nature of the international system. Korbel's world was essentially Hobbesian, updated and operationalized by Hans Morgenthau. The nation state was the unit of analysis, conflict and power the governing concepts, scarce resources the objective of political behavior. He described the world as it was rather than speculating on what it might become; a pathological international system functioned and, perhaps, was incapable of reform. In that world Soviet and Nazi conspirators, in the course of a stubborn and deceitful diplomatic struggle, sought to destroy Poland's independence or at the very least reduce Poland to impotency. Whatever the superficiality of appearances, ". . . there smouldered the fires of national mistrust, contempt, ideological hostility and territorial claims, eradicable conflicts of emotions, ideas, and interests that threatened to erupt at any moment."[25]

Transactions at two levels govern the diplomatic process; the official, formal, and courteous on the one hand, and the subterranean, covert, and destructive on the other. The former seem to point toward cooperative behavior but in fact represent illusion; the latter deal in duplicity and represent realism. Lest one should expect more of the West than, for example, of Nazi Germany or the Soviet Union, Korbel added the additional charge of ineptness in describing the Anglo-French record, an ineptness which, with the other qualities of their diplomacy, earned them the contempt and distrust of Eastern Europe.[26]

Korbel spent less time than one would expect on the subject of nationalism which he at times conceded was the key to so much of the history of the nineteenth and twentieth centuries. Nationalism was for him a matter of choice, a permanent, vital, and influential force for good *and* evil. Nationalism could, on the one hand, generate rich cultural legacies, unite a people in the common good, provide a barrier, however inadequate, to communism and stimulate a demand for diversification, for autonomy and independence, for polycentrism. Moreover, and reminiscent of Beneš, nationalism provided a stabilizing force in the international system when the maintenance of the sovereignty and integrity of each nation state was equated with order and justice. On the other hand, nationalism narrowed the vision of a people, bred ethnocentrism, cultivated hatred, and its logical outcome could be facsism or Nazism. Finally, it could balkanize and destabilize an area, and actually make it more vulnerable to communism.

He spent perhaps as little time as one would expect on the subject of religion and, indeed, broached the question almost exclusively in its political context. Religion was a beneficial stimulant in that it bred among the people a skepticism for monolithic political structures and a hatred of the arbitrary use of force which characterized totalitarian regimes. The church in Eastern Europe, led by the courage of its hierarchy, was a focus of dissent, a proponent of freedom and, by definition, an enemy of communism. To that extent, the church had earned his respect.

I have described Josef Korbel as an emigré and testified to his profound objection to communism but he was no crude Cold Warrior. Nevertheless, he indicted the Soviet Union, but not at any length, with the principal responsibility for the breakdown of the war-time Allied coalition, the creation of the Eastern Bloc, the rejection of the Marshall Plan, and for the deepening tensions of the bi-polar world.[27] Stalin was, therefore, primarily responsible for both the origins and the perpetuation of the Cold War. The work of the Revisionists did not impress him. While the United States and her allies were not devoid of error, they were guilty more of a weakness bordering on appeasement, of a lack of prescience and plan, and of a particularly debilitating tendency to fluctuate between firmness and irresolution, than immoral intent.[28]

Finally, no theme captures the essence of Josef Korbel more appropriately and fully than that of the role of the educated person, and especially the educated young person, in the history of Eastern Europe and particularly of Czechoslovakia. Universities and their students represented both present achievement and the hope of the future because, most fundamentally, they were the centers of dissent, the fountains of iconoclasm, and the focus of free thinking. They were, moreover, the residual energy of a nation and the basis from which came the will to fight oppression.[29] Korbel discovered in the students of Yugoslavia and Czechoslovakia a contempt for Marxist theory, dismay at its application, and resistance to its infiltration, dissemination, and adoption. University students were no less than the "vanguard of the democratic struggle for freedom," a haven for the nation's

progressive thinking.[30] Education itself was "the very *officina humani-tatis.*"[31] With considerable pride he presented the statistics of November 1947 when, in Czechoslovakia, seventy-four percent of the university students voted in their elections for democratic candidates and a mere twenty percent voted for the Communists.[32] These results, moreover, were predictable for the students were the indicator of the political mood of the country, its instinct, its pulse, and its temperature in matters of decency and wisdom.[33]

It follows, logically, that Korbel poured contempt and condemnation on those regimes which debased education by using it for indoctrination and propaganda. When the classroom became a barracks for political training, as it did in Communist Eastern Europe, education became a curse not a blessing. Under Gustáv Husák, for instance, "the nation became an intellectual and spiritual cemetery."[34] And this Josef Korbel could not forgive for above all else he respected scholars and scholarship, intellectuals and things of the intellect, and education and those who devote their lives to its well being.

NOTES

1. *An Ambassador's Report on Tito's Communism* (University of Denver Press, 1951). *The Communist Subversion of Czechoslovakia, 1938-1948: The Failure of Coexistence* (Princeton University Press, 1959). *Poland Between East and West: Soviet and German Diplomacy Toward Poland, 1919-1933* (Princeton University Press, 1963). *Detente in Europe: Real or Imaginary?* (Princeton University Press, 1972). *Twentieth Century Czechoslovakia: The Meanings of Its History* (Columbia University Press, 1977). His other book was *Danger in Kashmir* (Princeton University Press, 1954).

2. *Tito's Communism,* p. 182.

3. *Detente in Europe,* pp. 85-88.

4. *Twentieth Century Czechoslovakia,* p. 5.

5. Ibid., p. ix.

6. *Detente in Europe,* p. 143.

7. *Twentieth Century Czechoslovakia,* p. 5.

8. *Tito's Communism,* pp. 300-312.

9. *The Communist Subversion of Czechoslovakia,* p. 221.

10. *Twentieth Century Czechoslovakia,* p. 24.

11. Ibid., p. 145.

12. *The Communist Subversion of Czechoslovakia,* pp. 49-50, 129, 179.

13. Ibid., pp. 84, 114.
14. *Twentieth Century Czechoslovakia*, p. 216.
15. Ibid., p. 267.
16. *Tito's Communism*, p. 135.
17. Ibid., p. 224.
18. *Twentieth Century Czechoslovakia*, p. 271.
19. *Tito's Communism*, pp. 231, 307.
20. *Twentieth Century Czechoslovakia*, pp. 82–84.
21. Ibid., pp. 125, 148.
22. Ibid., p. 253.
23. Ibid., pp. 319–320.
24. *Poland Between East and West.*
25. Ibid., p. 103.
26. Ibid., p. 93.
27. *Tito's Communism*, p. 269.
28. *Detente in Europe*, pp. 27–29, 213.
29. *Tito's Communism*, p. 137.
30. *The Communist Subversion of Czechoslovakia*, pp. 160, 191, 230.
31. *Twentieth Century Czechoslovakia*, p. 9.
32. Ibid., p. 244.
33. Ibid., p. 280.
34. Ibid., p. 314.

Bruce Garver

THE CZECHOSLOVAK TRADITION: AN OVERVIEW[1]

I. INTRODUCTION

The Czechs and Slovaks, like all European peoples, cultivate and take great pride in a rich heritage from the past. In modern times, they have lived together in one state only since October 1918, when they established an independent Czechoslovak Republic after having helped destroy the Habsburg Monarchy. Thereafter one may speak of a Czechoslovak heritage, while recognizing that in discussing the past or present, one more properly speaks of a Czech heritage and a Slovak heritage that are occasionally identical, frequently similar, but more often than not different in important respects. Each nationality has its own distinctive culture, history, goals and language. Czech and Slovak are entirely separate Western Slavic languages but so closely related that a speaker of one has little or no difficulty in understanding a speaker of the other. Communication has, therefore, been no serious problem in a relationship more often characterized by compromise and cooperation than by conflict.

This chapter aims to survey the principal features of the heritage of the Czechs and Slovaks in a European context to reveal the similarities and differences between each heritage as well as the contributions each has made to the development of modern Europe. The author more often emphasizes the Czech rather than the Slovak heritage not out of any personal preference for the former but because it is the larger of the two nations, has had a somewhat more clearly defined past, and has exerted greater influence in Czechoslovak and European history.

The Czech and Slovak Relationship

Czechs and Slovaks lived under different governments and laws from the Hungarian conquest of Great Moravia during the early tenth century until 1526, when an Austrian Habsburg became king of Hungary and of the Czech kingdom of Bohemia. Even then, the Habsburgs did not destroy the autonomy of Bohemia until the Thirty Years War nor did they or their allies drive the Turks entirely out of Hungary until 1697. Though the dynastic union of 1526 continued until the dissolution of the Habsburg Monarchy in 1918, Czechs and Slovaks still remained subject to very different local and provincial administrations and to different cultural influences, German in the former case and Hungarian in the latter. Before 1918, therefore, close relationships between Czechs and Slovaks developed primarily in intellectual and religious matters and to a much lesser extent in commerce and politics. Such relationships, while often cordial, arose primarily from mutually intelligible languages, geographic proximity, joint advocacy of Slavic reciprocity, and, after 1848, similar political aspirations, including the advancement of national autonomy, civil liberties, social reform, and the use of native languages in schools and public life.

In the Czechoslovak Republic relations between Czechs and Slovaks became more extensive and intensive, thus increasing the incidence of misunderstanding and conflict, especially since the Czechs, by virtue of their greater numbers, wealth, and political experience, and often with the best of intentions, insisted upon being the senior partner. Slovaks especially resented the efforts of some Czechs to promote the fiction of a single Czechoslovak nationality, when in fact most Slovaks as well as Czechs viewed themselves as separate but closely related peoples. Both nonetheless tried to make the new Republic work, in part out of enlightened self-interest and in part because most Slovaks recognized their inability to go-it-alone and the fact that in Czechoslovakia they could develop their language and culture freely for the first time. Conflict between Czechs and Slovaks reached its height on the eve of the Second World War when Slovak separatists — with Nazi encouragement — set up the puppet state of Slovakia in March, 1939. The Slovaks who rose against that regime and the Nazis in August, 1944 not only redeemed themselves and fellow Slovaks in their own and in Allied eyes but helped persuade many Czechs to recognize the Slovaks as a distinct nationality entitled to a greater voice in postwar Czechoslovakia. In the early fifties, Slovaks were justly outraged when the Czechoslovak government and Communist party denounced leading Slovak Communist intellectuals as "bourgeous nationalists," incarcerating many and executing Vladimír Clementis, former Minister of Foreign Affairs, on trumped-up charges of "anti-state conspiracy." Relations between Czechs and

Slovaks improved from the early sixties through the short-lived Dubček era and the federalization of the country on October 28, 1969. This trend appears likely to continue, given the greater Czech recognition of Slovak autonomy, a mutual pride in past achievements, recently shared hopes and humiliations, and growing recognition that the continuing cooperation of both nationalities offers advantages to each that could not otherwise be obtained.

Geography and Early History

The Slavic ancestors of the Czechs and Slovaks began to settle in what is now Czechoslovakia sometime before the sixth century A.D., replacing in the West a German tribe, the Marcomanni and their predecessors, a Celtic tribe, the Boii, after whom the Romans named Bohemia. The Czechs (*Češi* or *Čechové*) took their name from Father Čech, who, according to legend, led them into Bohemia (*Čechy* in Czech).

As the westernmost Slavic people, the Czechs have been more involved with and influenced by the Germans than other Slavs. This relationship has been characterized by mutually beneficial interaction as well as by conflict, with Czechs generally having been more willing to accept German economic and cultural influence than German political preponderance. The Czechs have so tenaciously withstood German eastward expansion that the linguistic frontier between the Czechs and the Germans has moved no more than thirty miles in any direction during the past twelve centuries. The fact that Bohemia is surrounded by mountains as well as the fact that the Czechs organized a strong medieval kingdom helped them preserve their language and national identity. The Slovaks (*Slováci*) also inhabited very mountainous terrain and this may have helped them maintain their Slavic culture and language during ten centuries under Hungarian rule.[2] Slovakia and the Czech lands of Bohemia, Moravia and southeastern Silesia are of great strategic importance because of their rugged terrain and because they lie at the heart of Europe. For these reasons and because of their great agricultural, commercial and mineral wealth, they have often been subjected to foreign invasions and conquest. These lands have also long been centers of European commerce, standing astride the main trade routes between Western and Eastern Europe. From the Middle Ages to the present, this central location has in turn encouraged Czechs (and to a lesser extent Slovaks), to try to mediate between East and West, between Slavs and Germans, and between Orthodoxy and Catholicism.

II. SIMILAR CHARACTERISTICS OF THE CZECH AND SLOVAK HERITAGES

The Czech and the Slovak heritages resemble one another and those of certain other European nations in at least four important respects. First, both the Czechs and the Slovaks are small nations that in the past and present have often had to endure foreign domination. Second, like all Slavs, Czechs and Slovaks take pride in belonging to the largest family of nations in Europe and sharing in its rich and diverse cultural heritage. Third, Czechs and Slovaks, like other predominantly Roman Catholic or Protestant peoples of Eastern Europe, are decidedly Western in cultural and political orientation. Finally, both Czechs and Slovaks were almost exclusively peasant peoples without native aristocracies from the early seventeenth century until the advent of industrialization and urbanization two centuries later.

Small Nations

The Czech and Slovak experience in many respects resembles that of other small European nations which for centuries have endured foreign rule, had no native aristocracy, and been unable to maintain an unbroken literary development. Small nations of this sort include all those in Eastern Europe except the Poles and Hungarians and in Western Europe those nations that won their independence within the last century like the Norwegians and Irish, and those still seeking greater autonomy within larger nation states, e.g.: Basques, Bretons, Catalans, Scots and Welsh. The difficulties that small nations have encountered in trying to assert their independence in a Europe dominated by great powers were clearly delineated by the Czech philosopher and statesman, T. G. Masaryk (1850–1937), in *The Problem of the Small Nation* and *The Czech Question*. The great Czech literary critic, F. X. Šalda (1867–1937) defined the difficult task faced by artists, writers and composers from small nations in creating works that are true to national traditions and universal in appeal.

In the early Middle Ages, the Czechs, like the Serbs and the Bulgarians and other East Europeans under native dynasties, established kingdoms that centuries later were absorbed by larger empires. The medieval Czech Kingdom of Bohemia — which included Moravia, Lusatia, and Silesia as well as Bohemia — maintained its autonomy within the Holy Roman Empire from the twelfth century until the Thirty Years War (1618–1648). A native Czech dynasty, the Přemyslids, ruled Bohemia from the early tenth century until 1306, when King Václav III died without male heirs. Heads of the Luxembourg family, related to the Přemyslids by marriage, ruled in Prague from 1310 until 1437 as elected kings of Bohemia and as Holy Roman Emperors. Charles IV (1346–1378), son of John of Luxembourg and Eliška Přemyslovna, gave the medieval Czech kingdom a "Golden Age" of prosperity, peace, culture, and learning. For several decades after the martyrdom of John Hus in 1415, the Czech Hussites introduced extensive religious reforms, including Communion in both kinds for the laity,* and advanced the use of the Czech language in church and state. At the same time, they defeated every crusading army sent against them by the Empire and papacy. George of Poděbrady, a Hussite warrior and nobleman, who reigned from 1458 to 1471, was the last native Czech king. For the next century and a half, Czech nobles continued to dominate the Bohemian and Moravian Estates under elected kings from the Polish Jagiellonian dynasty until 1526 and thereafter under Austrian Habsburg kings, elected in 1526 but hereditary after 1547. During the Jagiellonian era, noblemen in the Czech lands gradually enserfed their peasants and reduced the privileges of many royal cities. Under the Habsburgs, German influence increased markedly in court and state administration. Meanwhile, German had already begun to replace Latin as the principal international language of Czech clergymen and scholars, while Czech popular culture had come to resemble the German in several respects, notably in cooking and in the use of beer as the staple alcoholic beverage. In every century, Czechs have benefited from close cultural and commercial contact with Germans, while at the same time working to maintain and enrich their own language and distinctive national heritage.

In order to maintain religious toleration and their traditional prerogatives, the predominantly Czech and Protestant Bohemian Estates revolted against the Habsburgs in 1618, only to be defeated at the White Mountain in 1620 by the armies of the Habsburg Emperor and

---

*Editor's note: "Communion in both kinds" refers to the offering of the chalice by a celebrant during Communion.

the Catholic League. After dealing harshly with all rebels, the Emperor, advised by the Jesuits, abolished Bohemian autonomy in 1627 and outlawed Protestantism, and began the gradual replacement of Czech by German as the official language of state administration. Only those Czech nobles who had loyally upheld Catholicism and the crown kept their titles, land, and political privileges. Meanwhile, foreign noblemen, largely German-speaking, who had fought for the Habsburgs acquired the lands of rebellious noblemen executed or exiled after the defeat at White Mountain. For the next 160 years, during "the era of darkness," Czech survived as the language primarily of peasants, parish priests and small tradesmen.

The Slovaks, by contrast, established no medieval kingdom, though most historians consider the ninth century Slavic state of Great Moravia to be a distant forerunner of modern Slovakia and Czechoslovakia. That Christian state, which encompassed territories now in Moravia, Slovakia and Lower Austria, fell to the Magyars in the early tenth century. They incorporated what is today Slovakia into the Kingdom of Hungary, while Moravia came under the control of the Czech principality of Bohemia. For the next millennium thereafter under Hungarian rule, Slovak peasants maintained their Slavic language and distinctive folk culture but adopted many Hungarian practices, notably in cuisine, music, and religious art and architecture. Slovak peasants usually resented their domination by Hungarian landlords and state officials but seldom came into conflict with Hungarian peasants or clergymen.

Slavic Reciprocity

Czechs and Slovaks, like all Slavs, have long taken pride in their Slavic heritage and as small nations have derived some confidence from belonging to the largest of the three great language and ethnic groups in Europe. That this pride was already evident among Czechs during the Middle Ages is best illustrated by Charles IV, who established in Prague the first university for Slavs and Germans and a monastery for monks practicing the Old Church Slavonic liturgy. He expressed pride in his Slavic ancestry and confidence in the future of the Slavs. During the early nineteenth century, Czechs and Slovaks, like other small Slavic peoples undergoing national revival, advocated solidarity and cooperation between all Slavs in cultural, commercial, and political matters. Such reciprocity in no way required the subordination of one Slavic nation to another or of all Slavic nations to Imperial Russia because Czechs and Slovaks continued to think of themselves first as Czechs and Slovaks and secondly as Slavs.[3] Among the earliest and foremost proponents of Slavic reciprocity were Josef Jungmann (1773–1847), Czech grammarian, poet and literary historian; Pavel Josef Šafařík (1795–1861), Slovak ethnographer; Ján Kollár (1793–1852), Slovak epic poet and Protestant pastor; Karl Jaromír Erben (1811–1870), Czech folklorist; and František Palacký (1798–1876), the great nineteenth-century Czech historian. All helped make Prague a leading world center of Slavic studies by the 1840s, a distinction it has never relinquished.

Western Orientation

The Czechs and Slovaks, along with the Poles, Slovenes, Croats, Hungarians, and Lusatian Sorbs, have maintained a decidedly Western political, social and cultural orientation since the tenth century. Though missionaries from Byzantium, including Saints Cyril and Methodius in 863, first brought Christianity to Bohemia and Great Moravia, by the late eleventh century Czechs and Slovaks had aligned themselves with Rome and begun to adopt Roman Catholic liturgy and institutions. Unlike most Orthodox Christian Slavs, they participated in and contributed to European culture and society from the Middle Ages through the Renaissance, Reformation and Counter-Reformation. For example, the great age of early modern Czech literature and science coincided with and resembled the achievements of the Elizabethan age in England. The nineteenth-century national revival among Czechs and Slovaks, as among other small European nations, in large part developed in response to the example and ideas of the Enlightenment, the French Revolution and Romanticism, and to the great transformation of society wrought by industrialization and urbanization. In style and content, Czech and Slovak arts, letters and music have been decidedly Western since the Middle Ages. This is evident not only in Gothic, Renaissance and Baroque arts and architecture but more recently in Czech and Slovak contributions to romanticism and realism in literature and to impressionism, expressionism and cubism in the arts.

During the Middle Ages, Czech and to a lesser degree Slovak society came to resemble that of feudal Western and Central Europe, a development facilitated by Czech participation in the Holy Roman Empire and by both the Czechs and Slovaks having earlier adopted Roman Catholicism and the Latin as opposed to the Cyrillic alphabet. From the mid-fifteenth century onward, the gradual imposition of serfdom upon Czech and Slovak peasants and the subjection of towns to royal authority paralleled similar developments elsewhere in those lands east of the Elbe and north of the Ottoman Empire.

From the tenth to the seventeenth centuries, Czech society differed markedly from that of the Slovak to the extent that a native Czech nobility shared in the administration of the Kingdom of Bohemia, whereas no Slovak nobility (some poorer gentry excepted), appears to have survived the Magyar conquest. From the 1620s onward the absence of a native landed aristocracy characterized Czech as well as Slovak society and differentiated these two predominantly peasant nations from the neighboring Germans, Hungarians and Poles. Noble families, primarily of German origin or inclination, largely controlled the politics, society and economy of the Czech lands until the eighteen sixties, just as Hungarian noblemen dominated most areas inhabited by Slovaks until 1918.

The Revolution in the Habsburg Monarchy led the Emperor to emancipate all peasants from feudal dues on September 7, 1848 and to replace the manorial system with a centralized imperial administration. The introduction of limited civil liberties and self-government in the eighteen sixties completed the status transformation from subjects to citizens and led to the immediate formation of Czech and Slovak political parties. Rapid industrialization and urbanization also furthered the growth of a large Czech middle class and industrial proletariat by the turn of the century. But, because of the backward economy and authoritarian government of Hungary, Slovaks remained an almost exclusively peasant people well into the twentieth century. The fact that both the Czechs and Slovaks entered the industrial era as predominantly peasant societies without native aristocracy and under foreign domination explains in large part their decidedly egalitarian and nationalistic politics as well as their advocacy of social justice and their cultivation of folk themes in literature, music, and the arts.

III. IMPORTANT ASPECTS OF THE
CZECH AND SLOVAK HERITAGES

Let us now examine in greater detail several specific aspects of the heritage of each nationality with a view to understanding past and present developments and some of the differences between each heritage. The legacy of the past will thus be revealed in the arts and letters, in religion, in agriculture and industry, in education and scholarship, and in politics. In all of these areas, Czechs and Slovaks have distinguished themselves and helped create a cherished national heritage.

Arts and Letters

Czechs and Slovaks look back with pride upon a rich cultural tradition, which is almost entirely Western in orientation. The Czechs, to a greater degree than the Slovaks, created a flourishing religious and secular art and literature from the late Middle Ages through the sixteenth century, with most works after the fourteenth century appearing in the vernacular as opposed to Latin. During the Thirty Years War, the Habsburgs and the Church, having associated the Czech language with rebellion and heresy, burned much of the secular and Protestant literature in Czech and placed higher and secondary education under Jesuit control. This campaign and the gradual replacement of Czech by German in state administration and in higher schools resulted in a situation where Czech became an increasingly colloquial language spoken largely by peasants, poorer tradesmen and parish priests.

The Czech intellectuals and clergymen of the later eighteenth century who began to revive Czech as a literary language chose to do so in large part upon the basis of Bohemian Czech and the surviving masterworks of fifteenth and sixteenth century prose. Their Slovak counterparts, however, decided that any standardized written Slovak language should closely resemble everyday speech. A majority supported Ľudevít Štúr (1815–1856) in basing their language upon the central Slovak dialect most intelligible to all Slovaks instead of the western Slovak dialect that more closely resembled Czech.[4] This decision formalized and accentuated the minor differences between the Czech and Slovak languages and facilitated the growth of a Slovak national consciousness. The fact that the Slovaks, unlike the Czechs, had neither a literary language nor an autonomous kingdom in medieval or early modern times explains in part the great influence of folk art and customs in defining their national identity and in conditioning the development of national arts and letters. This also helps explain why during the nineteenth century Slovak writers to a somewhat greater degree than Czechs utilized Slavic as opposed to peculiar national themes. Representative of this predilection were such famous works as the great epic poem *The Daughter of Slava* by Ján Kollár and *Hail,*

Slavs! by Samo Tomašik (1815–1887), the most popular of all hymns to Slavic solidarity.

During the nineteenth century, the Czechs became a cultured European nation in every sense of the word. The Slovaks, despite poverty and, after 1874, Magyarization, advanced toward the same goal. Most Czech and Slovak artists, writers, and composers looked to an international as well as a national audience and sought to create distinctively national works of art that would also have universal appeal. They aimed to strengthen national consciousness among all social strata and to make original contributions to European arts and letters. Most aspired to popularize higher culture and scholarship, thus helping a mass audience better appreciate both national and foreign artistic and intellectual achievements.

Many Czechs and Slovaks thought that fine art, literature and music ought to reflect and refine the best in popular or folk culture. Others, more often than not Czechs, stressed the need to overcome parochialism and excessive national pride by critically evaluating folk traditions and by promptly adopting the latest European styles and methods. As a rule, the first group included many Romantics, while in the more cosmopolitan second group, one often found symbolists in literature and impressionists or expressionists in the arts. These cosmopolitans may be viewed as the heirs of those Czechs and Slovaks who had for centuries adopted new European styles, beginning with French Gothic in the fourteenth century and including Italian and Spanish baroque in the seventeenth century, German romanticism and Russian epic poetry in the nineteenth century, French symbolism and impressionism at the turn of the century, and thereafter the international movements of cubism and surrealism. The Czech avant-garde of the twenties and early thirties especially emphasized the latter. Through the past ten centuries, Christian, and to a lesser extent Jewish and pagan Slavic, influences have usually been evident. The Czechs and Slovaks, in turn, have made original contributions to European arts and letters in the predominant styles of the time, as is evident in the medieval castles of central and southern Slovakia and the Gothic art of southern Bohemia, the Renaissance town halls and commercial buildings in the larger towns of Bohemia, Moravia, Slovakia, and more recently in Antonín Slavíček's contribution to impressionist art, in that of Josef Gočár and Pavel Janák to cubist architecture, and Viteslav Nezval and Vladislav Vančura to the literary and artistic avant-garde of the twenties and thirties.

Besides emphasizing national revival and Slavic solidarity and cultivating themes from folk arts and letters, Czech and Slovak literature of the nineteenth century was characterized by a strongly rational and empirical outlook, dating from the Enlightenment. Reinforcing this outlook was a Czech and Slovak concern for moral and social issues, first strongly manifested in the works of John Hus, the Hussites, and Petr Chelčický. This tradition, resurrected by František Palacký, and popularized by T. G. Masaryk, Jan Herben and the Slovak "Hlasists," may also be seen in the writings of the Czech Reformation, especially those of the Bohemian Brethren and J. A. Komenský, and subsequently in the works of the National Revival, including the ethnographical studies of Pavel Josef Šafařík, the novels of Božena Němcova, and the poetry and short stories of Jan Neruda. Karolina Světlá and later Tereza Nováková explored the subjection of women and the desirability of women's emancipation in essays and fiction. Julius Mrstík and Jakub Arbes exposed in novels, like the former's *Santa Lucia* and the latter's *Candidate for Existence,* many social problems arising from industrialization, urbanization, and an increasingly unequal distribution of income. A similarly penetrating realism and social criticism is evident in the novels of such recent Czech and Slovak writers as Ladislav Mňačko, Ludvík Vaculík, and Josef Škvorecký. Social satire in these and other works helps perpetuate a tradition well-established by that most popular of all Czech novels, Jaroslav Hašek's *The Good Soldier Švejk and His Fortunes in the World War.*

From the Romantic poet Karel Hynek Macha to the present, Czech and Slovak literature has preoccupied itself with subjective questions of love and morality as well as with the more objective questions of death and utility. Powerfully influenced by Byron and German Romanticism, Macha revived an artistic and individualistic Czech poetic tradition begun in early medieval oral tales and ballads, firmly established by the lyric poetry of the "Golden Age" of Charles IV, and reaffirmed by Czech baroque poets during the first decades of the "age of darkness."

A characteristic of Czech arts and letters of the past century and a half has been a powerful and graphic imagination that examines nature and society without illusions. That view of nature may be seen in the works of such landscape painters as the realist Antonín Chitussi (1847–1891), the impressionist Antonín Slavíček (1870–1910), Emil Filla (1882–1953), and Ladislav Čepelák (born 1924). A similar and more mystical imagination as well as new departures in style may be seen in such classic works as the symbolist poetry of Otakar Březina

(1868-1929) and the musical compositions of Leoš Janáček (1854-1928). An equally intense and very analytical imagination is evident in such recent creations as the poetry of Miroslav Holub, the plays of Václav Havel, the novels of Ludvík Vaculík, and in the films of Miloš Forman, Věra Chytilová, and Jiří Menzel. Contemporary works of art that take a hard, imaginative, and often skeptical view of man and his achievement include the graphics of Naděž Plíšková, the sculpture of Karel Nepraš, and the drawings of Jan Steklík. A softer but equally explicit imagination characterizes the work of contemporary Slovak artists like Viliam Chmel and Zolo Palugyay.

Among the most imaginative works of social criticism in any language are those of Karel Čapek (1890-1938), the Czech writer best known and most read in English-speaking countries. Among his most admired works at home are the collections of short stories titled *Tales from One Pocket* and *Tales from the Other Pocket* and the trilogy of novels published during the 1930s — *Hordubal, Meteor,* and *An Ordinary Life.* Abroad, Čapek is better known for his anti-utopian novel, *War with the Newts,* and his play of 1922, *R.U.R. (Rossum's Universal Robots),* that spoke of a world devastated by war and introduced the word "robot" into almost every language.[5] Both works deal primarily with contemporary problems but look back in small part to such legends as the Golem and Dr. Faustus and forward to some of the socially critical science fiction of today.[6] Čapek argues that unless man learns to master himself, he may destroy all civilization.

Since the early nineteenth century, Czech and Slovak artists and intellectuals have participated in and influenced the social and political development of their nations in at least four important respects — as politicians, as ideologues, as social critics, and as leaders of cultural and patriotic organizations. First, from František Palacký and T. G. Masaryk to contemporaries like Václav Havel and Pavel Kohout among Czechs; and from Ján Hollý (1759-1849), and Svetozár Hurban Vajanský (1847-1916), to Laco Novomeský and Ladislav Mňačko among Slovaks; intellectuals have served as political spokesmen, besides advancing the arts, letters or scholarship. Second, certain works like Palacký's *History of the Czech People,* Božena Němcová's (1820-1862) novel *The Grandmother,* and Bedřich Smetana's opera *The Brandenburgers in Bohemia,* have helped define Czech national aspirations as well as delineate the national character and heritage. Comparable works by Slovaks include Samo Chalupka's (1812-1883) epic poem *Strike him dead!,* Ján Levoslav Bella's vocal and instrumental piece *The Marriage of Janošík**, and Ladislav Nádaši-Jégé's (1866–1940) novel *Adam Šangala: A Historical Portrait of the Seventeenth Century*. Third, Slovak and especially Czech intellectuals have critically evaluated and tried to improve all aspects of their national experience, including politics, society and the arts. Fourth, Czech and Slovak cultural, educational and patriotic organizations have performed important political tasks. Cases in point are the success of theaters, orchestras and lending libraries in helping arouse national consciousness during the formative years of the national revival, and the subsequent efforts of organizations like the Slovak Foundation (*Matice Slovenská*) and the Czech *Sokol*, a patriotic fraternal and physical fitness organization, to advance national solidarity and moral and intellectual development.

Through music, Czechs first won international recognition in the nineteenth century as a distinct cultural nationality. The much acclaimed orchestral works and operas of the Czechs Bedřich Smetana (1824–1884) and Antonín Dvořák (1841–1904) and the Slovak Ján Levoslav Bella (1843–1936), rest on a strong and centuries old heritage of folk and religious music. Folk music has enriched both popular and classical music during the past two centuries. Religious music developed out of the Slavic liturgy used by Czech and Slovak priests as late as the eleventh century and from the Gregorian chant introduced with the Latin liturgy from the tenth century onwards. Hymn singing, like Communion for the laity, reflected the popular outlook and goals of the Hussites. On several occasions, imperial and papal armies fled at the sound of Hussite armies singing favorite hymns such as *You Who are Warriors of God.*

Deprived of religious and civil liberty during centuries of foreign rule, Czechs and Slovaks often proclaimed their aspirations, joys, and sorrows and helped preserve their national consciousness through music. Modern composers like the Czechs Leoš Janáček and Bohuslav Martinů (1890–1959), and the Slovak Eugen Suchoň (born in 1908), have carried on this tradition.

Religion

Religion has powerfully conditioned Czech and Slovak cultural, intellectual and political development through the centuries. Roughly three

---

*The legendary folk hero Janošík is a Slovak Robin Hood.

in every four Czechs and Slovaks are Roman Catholics, though among neither people has Roman Catholicism so strongly influenced politics and society as among the Poles or the Irish. The Czechs and Slovaks are the only Slavic peoples with large Protestant minorities — roughly ten percent and sixteen percent respectively — and a strong Protestant intellectual tradition.[7] Both peoples, especially the Czechs, have in the past initiated and supported reform movements within Roman Catholicism from the Hussite movement of the fifteenth century to the Catholic Modernism condemned by Pope Pius X in his 1907 encyclical *Pascendi Dominici Gregis.*

Initiated by John Hus (1370–1415), and contemporaries, and influenced by the teachings of John Wycliffe (1328–1384), and Czech reformers like Matthew of Janov (1353–1393), the Hussite movement aimed primarily to reform and purify the Church and to encourage rich and poor alike to lead a Christian life. It also tried to curb the abuse of wealth and privilege, helped stimulate Czech national consciousness and encouraged the use of the Czech language in churches, schools and public life. After the Council of Constance burned Hus for heresy in 1415, his followers, including Jakoubek of Stribro (1372–1429), and John Rokycana (1395–1471), led a popular uprising against attempts by the Pope and the Emperor to reassert their control over the Church and state in Bohemia. Hussite wagon armies under the blind general John Žižka (1360–1424) and his successors defeated every invading force hurled against them. Hussite aims, exemplified by the Four Articles of Prague of 1420, called for the laity to receive wine as well as bread at communion, freedom for ordained priests to preach the word of God, a return by the clergy to "apostolic poverty," and the punishment by appropriate authorities of sin among all clergy and laity.[8] Hussite moderates under Rokycana came into conflict with a militant left wing that believed all Christians not only to be equal in the eyes of God but entitled to social and political equality here on earth. The victory of the moderate over the radical Hussites in 1434 at the battle of Lipany opened the way to reconciliation with the Emperor and the Pope, in which the latter recognized an autonomous Hussite or Utraquist church and confirmed John Rokycana as archbishop of Prague. Meanwhile, inspired by the preaching and example of Petr Chelčický (1390–1460), devout Czechs who had refused to take arms in defense of their Hussite faith began organizing in small groups to live a disciplined Christian life largely apart from what they considered to be an irreparably corrupt society. In 1467, several such groups formed a separate church, the Unity of the Brethren (*Jednota bratrská* or *Unitas fratrum*), that grew steadily in the Czech lands until proscribed in 1627. Thereafter in exile, the Brethren, often better known as Moravians, established congregations and missions worldwide, especially in Germany and North America.

Given the presence of Utraquists and the Brethren in the Czech lands, Protestantism made rapid headway during the sixteenth century, becoming by 1609 the religion of roughly two in every three Czechs and Slovaks. Protestants produced the first Czech translation of the Old and New Testaments, publishing the latter in 1595 and the former at Kralice between 1579 and 1593. The Kralice Bible and the hymnbooks of the Brethren, Hussites and Lutherans helped inspire a flourishing Protestant devotional and theological literature. After the imperial proscription of Protestantism in 1627, most Czech Protestants converted to Catholicism, while others either fled abroad or survived underground until Emperor Joseph II granted toleration to Protestants and Jews in 1781. Attempts to forcibly re-Catholicize the Protestant Slovaks, begun in the 1670s, achieved similar results especially after the authorities burned churches and sentenced many ministers to service on galleys in the Mediterranean. The first Slovak translation of the Bible nonetheless appeared in 1722; the works of Matej Bél (1684–1749), and Daniel Krman (1663–1740), were based upon the Kralice text.

Roman Catholicism among the Czechs and Slovaks dates back more than eleven centuries. Its association with the autonomous medieval Czech state began with the designation of the martyred Prince Václav (921–929), as the patron saint of Bohemia. Papal establishment of a bishopric in Prague in 973, a bishopric in Olomouc, Moravia, after 1060, and an archbishopric in Prague in 1343, testified and contributed to the growing Czech control of the Church in the Czech lands. After the Thirty Years War, Czechs continued to serve as parish priests but held very few higher administrative offices. Some, like Bohuslav Balbín (1621–1688), helped keep Czech alive in churches and in parish schools during the "era of darkness." Beginning with Josef Dobrovský (1753–1829), author of the first history of the Czech language and the first modern Czech-German dictionary, Czech Catholic priests and laymen served as leaders and as foot soldiers in the national revival. Nonetheless, the close association of Roman Catholicism with authoritarian Habsburg rule and the control of higher Church offices by Germans led many Czechs to regard the Church as an obstacle to national emancipation and social reform. Anticlericalism characterized Czech Social Democratic as well as Liberal middle class politics from the 1870s to 1914, especially after the Church, beginning with Pope Pius IX (1846–1878), assumed an increasingly reactionary posture toward civil and ecclesiastical affairs. Czech anticlerical Catholics and free-thinkers resembled in many respects their French and Italian counterparts and far outnumbered Czech Protestants and Jews. Though all anticlerical Czechs and Slovaks strongly advocated separation of church and state, none sought to restrict public religious worship or abolish private religious instruction.

From 1848 to 1918, Slovak Catholic and Lutheran clergymen helped lead the struggle for equal rights for all nationalities in the Kingdom of Hungary. Lutheran pastors like Jan Kollár and Ľudevit Štúr who led the national revival set the pace for Josef Miloslav Hurban (1817–1888) and Michal Miloslav Hodža (1811–1879), who numbered among the principal Slovak political spokesmen from the Revolution of 1848 through the 1860s. Andrej Hlinka (1864–1938), a Catholic priest who courageously demonstrated against Magyarization, turned the Slovak Peoples party into the largest Slovak party in Hungary before 1914, and up to 1938 in the Czechoslovak Republic, where it drew from one in three to one in four Slovak votes. By contrast, Czech clerical parties in the elections of 1907 and 1911 won less than one in eight Czech votes in Bohemia but an average of one in three Czech votes in Moravia. The fact that Catholicism and Protestantism were proportionally much stronger in Moravia than in Bohemia may in part be explained by the greater industrialization and urbanization of the province of Bohemia.

Most Czech and Slovak Catholic clergymen stood up for the Habsburg Monarchy until its disintegration appeared inevitable. Thereafter most clergymen supported the Czechoslovak Republic and tried to make its democratic representative institutions work, if only out of enlightened self-interest. Such priests also helped arrange the *modus vivendi* between Czechoslovakia and the Vatican in January 1928 that ended formal papal disapproval of the new state and its policies. During the Second World War, Monsignor Jan Šrámek, founder and long-time chairman of the Czech Catholic People's party, served under President Edward Beneš as prime minister of the Czechoslovak government in exile in London.

Since 1781, the Protestant minority among Czechs and Slovaks has exercised influence proportionately much greater than its numbers. This is evident in the many Protestant leaders of national revivals, of efforts to achieve political autonomy and social reform, and of the struggle for Czechoslovak independence during the First World War. In contrast to Hlinka's Slovak People's Party that sought the greatest possible autonomy for Slovakia within Czechoslovakia, most Slovak Protestants, regardless of social class, supported the Republic, despite reservations concerning the excessive centralization of governmental authority in Prague. Many Slovak Protestant intellectuals rose to prominence in governmental administration or in the larger Czechoslovak political parties, including Milan Hodža of the Agrarians and Ivan Dérer of the Social Democrats.

From the Middle Ages to the Second World War, Jewish communities in the Czech and Slovak lands contributed substantially to the development of culture, trade and urban society. Prague, Brno, and Bratislava were among the earliest and most influential European centers of Jewish commercial, cultural and intellectual life. Prague, for example, has the oldest surviving synagogue in Central and Eastern Europe, dating back to 1270, and also has a Jewish cemetery where gravestones erected in the tenth century still stand. Prague has been renowned since the early sixteenth century for its many Jewish writers and publishers. The Habsburg extension of toleration in 1781, and in 1860 of full civil rights to Jews and Protestants enabled men of both faiths in Bohemia, Moravia and Slovakia to advance rapidly in business and the professions. Up to 1918, most Jews in the Czech lands spoke German as their principal language and identified themselves with the politically dominant German minority. Likewise, Jews residing in predominantly Slovak areas usually chose to associate with the ruling Hungarian as opposed to the subject Slovak inhabitants. In Czechoslovakia after 1918, Jews increasingly declared themselves to be of Czech or Czechoslovak nationality, especially after the Nazi takeover of Germany in 1933.[9]

In comparison to other peoples of Central and Eastern Europe where large Jewish minorities used to dwell, the Czechs gave very little support to anti-Semitic movements. Why? First and most importantly, the strongly liberal and reform-minded orientation of Czech politics after 1848 precluded successful demagogic appeals to racial or religious prejudices. Second, because the Czechs competed as well against the Jews as against the Germans in all trades and professions so that Jews usually aroused little popular envy or resentment and were less likely to be made scapegoats by political demagogues. For example, Czech banks, businesses, and agrarian cooperatives and savings and loan associations were so powerful that Jews did not dominate retail trade or money lending in Bohemia and Moravia as they did in some areas of Eastern Europe. Third, many Czechs, among them Masaryk and the Progressives, contended that the Czechs, having long suffered discrimination in the Habsburg Monarchy, risked discrediting themselves and losing their moral posture as a justly aggrieved minority if they tried to discriminate against any other minority, including the Jews. Instead, Czechs ought to make common cause with non-German minorities everywhere in the Monarchy. Czechs, to be sure, on occasion boycotted Jewish merchants or gave preference to Czechs over Jews in hiring for private employment. They were especially critical of those Jews who endorsed German nationalist parties or Austro-Hungarian imperialism in the Balkans. But before 1914, less than one-half of one per cent of Czech voters ever cast ballots for candidates from avowedly anti-Semitic parties.

In the first Czechoslovak Republic, Czech and Slovak politics remained decidedly left of center with continuing emphasis upon civil liberties and social reform. Czechoslovak support for representative government and minority rights also contrasted favorably with the policies of neighboring East European states and with those of Germany after 1932. An extremely right-wing and anti-Semitic Czech party emerged for the first time in the nineteen twenties, but this Fascist National Front had incompetent leadership and virtually no popular following. Under Karel Kramář, Czechoslovak National Democracy, known for its intense Czech nationalism and hostility to Germans and German Jews, moved further to the right as its share of all votes cast declined from 6.3% in 1920 to 5.6% in 1935. The clerical Slovak People's party became increasingly critical of Czechs, Protestants, and Jews, but won no more than 30% of the vote in Slovakia and 6.9% nationwide. Unlike the Sudeten German party, formed in 1935, none of these parties seriously threatened the representative government or political stability of Czechoslovakia.

After the Germans established the Protectorate of Bohemia and Moravia in March 1939, a small minority of Czechs helped the Nazis implement anti-Semitic legislation. The puppet state of Slovakia, with Nazi encouragement, turned savagely upon its Jewish citizens just as it obligingly declared war on the Soviet Union and the United States. Today, most Czechs and Slovaks prefer to remember those occasions in which they resisted Nazism rather than discuss the many instances in which they either passively accepted or less often actively collaborated with Nazi authority. Despite the fact that many Jewish intellectuals played a prominent part in the Communist takeover of Czechoslovakia in 1948, the party's purge of alleged anti-state conspirators in the early fifties took on strongly anti-Semitic overtones. Today, primarily as a result of the Nazi-imposed "final solution," the Jews of Czechoslovakia are a small and not very influential minority, like those of all Soviet bloc countries except the Soviet Union.

Agriculture and Industry

From the early Middle Ages to the present, Czechs have been among the most prosperous and industrious people in Central and Eastern Europe. According to data which is fragmentary before 1860 but complete thereafter, the Czech lands of Bohemia, Moravia and Silesia have long enjoyed prosperity in agriculture, mining and manufacturing to a much greater degree than most parts of Eastern Europe and are comparable to Germany, Denmark and Sweden. Medieval and early modern Bohemia enjoyed close commercial ties with other economically advanced principalities of the Holy Roman Empire and was noted for the business districts in its royal cities and for the mining and coining of silver. From the late sixteenth century to the present, Czechs have produced and exported high quality glass, pottery, and musical instruments. The manufacture of porcelain, clocks, and textiles dates from the early eighteenth century.

The prosperity of the Kingdom of Bohemia since medieval times can in large part be attributed to its abundant raw materials and rich farmland as well as to its having been closely tied to the economy of the Holy Roman Empire. Heavy investment of foreign and Czech capital along with the industriousness and intelligence of the Czech people have also contributed to this prosperity. Czech farmers, for example, have been noted for their tenacity and productivity not only in the Haná of Moravia or the Elbe valley of Bohemia but on the prairies and Great Plains of the United States. A capacity for skilled labor and hard work has long characterized Czech and Slovak industrial workers as well.

From 1860 to 1918, the Czech lands were among the most intensively farmed and highly industrialized regions of the Habsburg Monarchy. In industrial rankings the Czechs were first in railway mileage, coal mining, and in the production of iron and steel, chemicals, paper, textiles, glass, armaments, and industrial machinery. In agriculture, Czechs accumulated capital and used it to finance a rapid industrialization that began in the 1860s with food processing industries like brewing and beet sugar refining, and culminated after 1890 in the manufacture of chemicals, electrical goods, precision machine tools, and heavy transportation equipment. Meanwhile, the "national industry" of growing and refining beet sugar continued to expand as Czechs, after inventing better beet slicers and raspers, achieved European pre-eminence in beet sugar production and in the manufacture of beet processing machinery. From 1918 to 1938, the Czechoslovak Republic ranked seventh or eighth among nations of the world in the total value of its industrial production. Since 1948, a decline in the quality and efficiency of Czechoslovak manufacturing and agriculture relative to West European and American standards may in part be attributed to slavish adoption of Soviet economic models and procedures, especially in regard to central planning and extraordinarily large investments in heavy industry.

Until recently, the Slovaks have engaged little in manufacturing. But, like the Czechs, they have mined and refined ores for centuries, especially in mountainous south central Slovakia around Banská Štiavnica and Kremnica. Many Slovak immigrants in the United States kept up this tradition by working in the coal mines and steel mills of Pennsylvania and Ohio. Given the comparative economic backwardness of nineteenth century Hungary, Slovaks perpetuated centuries-old folk art and handicraft industries to a greater extent than did the Czechs or Germans. Slovaks continue to produce folk art and handicrafts of great beauty and traditional design; the demand for such items having grown rapidly with the recent revival of tourism. Since 1945, the rapid industrialization of Slovakia has helped to create social and economic conditions approximating those in Bohemia and Moravia. Postwar developments include the manufacturing of steel at Košice and of heavy transportation equipment in Martin.

Science and Technology

Czech prowess in industry, agriculture and technology has owed much to the achievements of Czech and other European scientists and inventors in the Czech lands since the late sixteenth century. Scientific advances in chemistry during the sixteenth century helped improve brewing, metallurgy and assying ore in Bohemia and Slovakia and was reported by the contemporary publication of works by two Germans in Bohemia, Georgius Agricola (1494–1555) and Johannes Mathesius (1504–1565). In order to establish higher standards for pharmaceutical work in light of new knowledge, the Czech Adam Zalužanský of Zalužany (1556–1613) published his classic *Rules of Pharmacy* in 1591. During the reign of Rudolf II, Holy Roman Emperor and King of Bohemia (1576–1612), Prague became a leading European center for scientific experimentation and the world center for astronomical research and publication. There, from 1599 till his death in 1601, the Danish astronomer Tycho Brahe (1546–1601) made the observations on which Johannes Kepler (1571–1630), residing in Prague from 1600 to 1612, based his renowned treatise on *The Laws of Planetary Motion*. Preceding and contributing to the discoveries of Brahe and Kepler was the work of Czech astronomers Tadeáš Hajek (1525–1660) in Prague and Cyprián Lvovicky (1514–1574) in Nürnberg.

Scientific research in Bohemia and Moravia declined markedly during the era of darkness and did not again achieve European recognition until the early nineteenth century, following establishment of the Royal Bohemian Academy of Sciences in 1771, the revival of literary Czech, and the advent of religious toleration. Among outstanding early nineteenth-century Czech scientists, the best known are Jan Evangelista Purkyně (1787–1869), pioneer physiologist and discoverer of the "Purkyně effect," and Jakub Kulik (1793–1863), who drew up comprehensive tables showing decompositions into prime numbers. Bernard Bolzano (1781–1848), born in Prague of Italian parents, won recognition for work in mathematical logic and analysis, including formulation of the Bolzano-Cauchy condition for "the convergence of series." Later nineteenth-century Czech scientists included František Tilšer (1825–1913), who published important works on descriptive geometry and "the theory of the illumination of surfaces," and Antonín Bělohoubek (1845–1910), famed for his scientific studies of yeast and for founding the Research Institute of Fermentation Chemistry in 1887. The publication from 1880 to 1909 of the twenty-eight volume Czech language *Otto Encyclopaedia* testified to Czech success in systematizing and popularizing as well as advancing scientific and general knowledge. At that time, this encyclopedia was the largest in any Slavic language and remained for decades a standard reference work throughout the Slavic-speaking world.

During the nineteenth century as in centuries before, Slovaks in Hungary had few opportunities to acquire higher education or scientific training in their own language. Among those who overcame such obstacles and helped advance science and technology were Aurel Stodola (1859–1942), discoverer of the theory and the construction techniques of steam turbines, and Milan Rastislav Štefánik (1880–1919), an astronomer and aviator, who during the First World War became the leading Slovak in the struggle for Czechoslovak independence.[11]

Czech scientific research and technological innovation — as well as heavy investments of foreign and Czech capital facilitated the rapid industrialization of the Czech lands in the later nineteenth and early twentieth centuries, especially in the chemical, machine tool, electrical, and automotive industries that Czechs helped introduce to Central and Eastern Europe. Among the most famous nineteenth-century Czech inventors and engineers was Josef Ludvík Ressel (1793–1857) from Chrudim, who invented the screw propeller for steamship propulsion. Franz Anton Gerstner (1793–1840), born in Prague, from 1829 to 1832 supervised the construction between České Budějovice and Linz of the first horse railway line in continental Europe and from 1834 to 1837 oversaw the building of the first railway in Russia. Important Czech inventors and entrepreneurs of the later nineteenth century included Václav Laurin and Václav Klement of Mlada Boleslav, pioneers in Austria-Hungary in the design and manufacture of motorcycles and automobiles and František Krizik (1847–1941), an electrical wizard, pioneer builder of trolley lines, and founder in Prague-Karlín of one of the first firms in the world to mass produce electric lights and generators. Among the Czech iron and steel magnates who most successfully applied machine technology and mass production was Emil Škoda (1839–1900), of Plzeň, who excelled in manufacturing machine tools, arms, and locomotives. Tomáš Baťa established a shoe factory in Zlin, Moravia, in 1894. He studied the mass production of shoes in Massachusetts and had built his firm into the largest of its kind in Europe by the 1920s. After 1918, Czech automotive manufacturers helped set the world pace in innovative and imaginative design. For example, Tatra of Kopřivnice, Moravia, introduced aircooled rear mounted engines in 1923 and an aerodynamic "streamlined" design in 1936, while Laurin and Klement, absorbed by Škoda in 1925, was among the first European firms to begin assembly line production. During the twenties and thirties, Czechoslovak racing car designers and drivers gave international competitors a good run for the money.[12] Building railway equipment is another industry in which Czechs and Slovaks have excelled. Among the most powerful, efficient and handsome dual-purpose steam locomotives ever run in Europe were the Czechoslovak Class 475 and 198 "Mountains" built by Škoda from 1947 to 1955. In 1953, with its class E499 Bo-Bo units, Škoda became the first firm in Eastern Europe to introduce high-speed electrically-driven locomotives based on the pioneering Swiss "all-adhesion" design of 1944. After the Second World War, Czechoslovak firms also took the lead in

Eastern Europe in producing powerful streamlined trolley cars, using patents from the proven American PCC prototype of 1934.

Education and Scholarship

Czech and Slovak scientific and technological achievements reflect the Czech love of learning noted by foreign observers from the Middle Ages to the present. That trait has been evident in the literacy of the many as well as in the intellectual distinction of the few. Czechs have prided themselves on having established the Charles University in Prague in 1348, the first university created east of the Rhine and north of the Alps. The Czech Hussites encouraged popular elementary education as one means of helping everyone understand and apply the teachings of Christ. The best known Czech advocate of universal education and pedagogical reform is John Amos Comenius (Jan Amos Komenský 1592–1670), educator, bishop of the Unity of Brethren and a leader of Czech Protestant exiles during the Thirty Years War. Comenius not only advocated universal education as a means of reducing religious and political strife but devised improved teaching methods and devices like illustrated primers to help achieve that goal.

Under the terms of the imperial school law of 1869, Czechs built up a comprehensive system of public elementary and secondary schools that largely eliminated illiteracy within one generation and that powerfully stimulated Czech national consciousness and patriotism. The Prague Technical Institute, established in 1869 with instruction in Czech, encouraged work in the theoretical and applied sciences. Tens of thousands of graduates contributed greatly to the rapid development of industry in the Czech lands, while some faculty and students rose to leadership in the liberal Young Czech Party and its successor parties. Finally, with the division in 1882 of the Charles-Ferdinand University in Prague into separate Czech and German universities, Czechs celebrated their having established an excellent system of Czech schools at every instructional level. Until the dissolution of the Habsburg Monarchy in 1918, German opposition, sanctioned by the Emperor, kept Czechs from founding a Czech language university in Moravia, just as the Hungarians prevented the Slovaks from opening any institutions of higher education or operating more than two secondary schools with instruction in Slovak. From its inception, Czechoslovakia has maintained a higher rate of literacy than any Slavic country. In addition, the country has had universities and research facilities where scholars have achieved international distinction in all fields.

Politics

Czechs and Slovaks have a less well defined political heritage and have had a less independent political development than many larger European nations. Neither Czechs nor Slovaks enjoyed political independence before the years when they, like other European peoples, established themselves as modern political nations. But the years from the French Revolution to the 1890s saw rapid industrialization and urbanization facilitate the emergence of a national consciousness, accentuate the differentiation between social classes, accelerate the growth of political parties, and stimulate popular demands for universal male suffrage, extended civil liberties, and greater equality of opportunity.

Czechs and Slovaks, in the absence of any national state of their own until 1918, identified the political nation with the people and very little with any past or present state. But since 1918 they have identified themselves and their political aspirations with great achievements of the past, with the lands in which they have lived for nearly fourteen centuries, and with certain universal moral principles that might serve as guides to political conduct. Specific characteristics of Czech and Slovak political development include successful self-help and cooperative endeavors and political parties which have been primarily popular in inspiration, disciplined in organization, and pragmatic and peaceful in action.

Czech intellectuals and politicians of the nineteenth century revived many of the symbols of the old Kingdom of Bohemia, while most middle-class parties, including both the Old and the Young Czechs, based their liberal and nationalistic programs on Bohemian state rights or those rights that the autonomous Kingdom of Bohemia had enjoyed before 1627. The Slovaks, having had no comparable state, looked to their distinctive folk heritage for inspiration and symbols and based their claims for political autonomy almost exclusively upon the natural rights of man. In 1918, due to the overwhelming Czech preponderance in wealth and numbers, as well as the symbolic importance of historic Bohemia in Czech political life, the new Czechoslovak state adopted many Bohemian symbols as its own. For example, all presidents of the Republic have lived in the Prague Castle, the traditional residence of Bohemian kings, while the Czech lion, symbol of Bohemian sovereignty, figures more prominently than any other element in the Czechoslovak coat of arms.

Czechs, despite their emphasis on Bohemian state rights, have, like the Slovaks, based their claims for national autonomy and civil liberties primarily upon the natural rights of man. They also clearly distinguished between the nation and the state, defining the former in terms of all persons speaking Czech and identifying themselves as Czechs. The nation, whether Czech or Slovak, shared a common language and national identity and a common past and had lived for more than a millennium in the same clearly defined geographical area. Intellectuals also sought to justify the existence of the nation by identifying certain principles or ideals by which national politics might be guided or inspired — not only to help justify the nation in foreign eyes, but to define a national purpose or set of national objectives. Because of the importance of the people, of place, of the past and of political principles in nineteenth and twentieth century Czech and Slovak politics and their national heritages, a word about each is in order.

Identification of the nation with the people during the era of national revival came naturally enough to predominantly peasant peoples without a state of their own. The Habsburg Monarchy was hostile or at best indifferent to the political aspirations of its subject nationalities, whereas the Kingdom of Bohemia as well as Great Moravia appeared to be a relic of the distant past. František Palacký, the leading Czech politician and historian during the period 1848 to 1871, titled his magnum opus the *History of the Czech People* as opposed to "the history of Bohemia" and identified the popular Hussite movement, not the medieval Kingdom of Bohemia, as the high point of the Czech past. In both cases, the people took precedence over the state as the basis for defining the political nation.

Czech and Slovak national consciousness and national politics have long been strongly identified with place. Folk customs, legends, and great events from the past are all intimately associated with particular places, including castle ruins and small villages as well as great rivers and cities. The best examples of this are the Czech and Slovak national anthems, like those of most European nations written during the early to mid-nineteenth century. The Czech anthem, "Where is my home?" finds a home in all of the beautiful lands long inhabited by Czechs. The Slovak anthem, "Lightning in the Tatras," symbolizes the awakening of the Slovaks that takes place in the high Tatras of central Slovakia, the tallest mountains in the Carpathian range which have long been a symbol for majesty and strength. Other examples include Smetana's internationally acclaimed symphonic poem *My country* and the series of poems by Samo Tomašik on Slovak castle ruins. Regional as well as national consciousness has grown out of a strong sense of place. This is still evident in the preservation of local folk costumes and customs and in the pride in the past shown by particular regions or cities. In such cases, Czechs as well as Slovaks usually emphasize the extent to which place or region contributed to the making or to the welfare of the entire nation.

Czechs and Slovaks have looked to the past to provide a sense of national identity and political purpose and also to find a source of national pride and confidence. The national revival of the nineteenth century as well as the Hussite movement of the fifteenth century, has provided all of this to Czechs. Slovaks and Czechs alike have derived great pride and confidence from the fact that their nation has survived many centuries of foreign domination. In the case of each nation, shared experiences of the recent as well as the distant past have strengthened national consciousness and in recent decades formed some sense of Czechoslovak — as opposed to Czech or Slovak — solidarity.

The Czech and Slovak achievement of distinctive national cultures and of international recognition as cultured nations should be viewed as part of the larger effort by both nations during the nineteenth century to achieve greater economic prosperity and political and cultural autonomy within the framework of the Habsburg Monarchy. Advances by both nations in all areas occurred despite severe political setbacks in 1851, during the early eighteen seventies and after 1908. Through thick and thin, leaders of both nations have never ceased to express confidence in the survival and progress of their respective nations. Most Czechs and many Slovaks have considered the birth of the Czechoslovak Republic to be the logical, desirable, and crowning achievement of more than a century of national revival. The authoritarian Habsburg Monarchy had failed in trying to retard or control this revival and exercised at best superficial influence upon its success. In 1867, František Palacký accurately and prophetically defined the relationship between the Monarchy and its Slavic subjects: "We (Slavs) were here before the Habsburgs and shall be here after they are gone."

The Czechs and Slovaks both began to look to certain contemporary principles or ideals as well as to the past for political guidance and confidence during their national revivals. Principles like the natural rights of man not only justified efforts to achieve national autonomy but gave each small nation a sense of belonging to and participating in a movement or struggle much larger than the nation itself. Czechs and Slovaks derived confidence from their belief that European society as a whole was moving toward greater democracy and equality and that the authoritarian Habsburg Monarchy would have to reform in response to such trends or it would perish. In this respect, Czechs and Slovaks reflected the views of other small European nations during the nineteenth and early twentieth centuries. Among Czechs, T. G. Masaryk, the principal architect of Czechoslovak independence, most persuasively argued that political aims and tactics ought to be based on principles like work, humanity, justice, and the search for truth. A small nation, or for that matter any nation, would not long endure unless it were dedicated to universal goals.

During the past century and a half, Czechs and Slovaks, particularly the former, have excelled at practical self-help and at cooperative endeavor, whether this be in politics, in producers' and consumers' cooperatives, in manufacturing, in publishing or in urban and rural savings and loan societies. With most banking and higher offices of government dominated by foreigners until 1918, many Czechs and Slovaks used such self-help and cooperation to advance their material, political, and cultural well-being. During the later nineteenth century, this activity was more evident among Czechs than among Slovaks, because Hungarian law prohibited many cultural activities on the grounds that these disguised political agitation. The Hungarians, for example, abolished the Slovak Foundation in 1875 and confiscated its assets, whereas the comparable Czech organizations grew steadily from 1860 to 1914.

The hundreds of thousands of Czechs and Slovaks who emigrated from 1848 to 1914 took their cooperative, cultural and patriotic organizations with them. The Czech minority of over 100,000 in turn-of-the-century Vienna maintained private Czech language schools, newspapers, theaters, banks, savings and loan associations and various cultural organizations. Slovak and Czech immigrants in the United States established similar institutions, the Slovaks especially discovering in the United States much greater opportunity for collective and individual enterprise than they had heretofore known. Hence it is not surprising that a free Slovak press and patriotic institutions flourished in the United States.

Czech politics from the 1890s to 1914 in Austria-Hungary as well as Czechoslovak politics since 1918 have been characterized by competition between Czech or Czechoslovak parties representing various ideological and economic or class interests as well as by conflict between the Czechs and Germans and between Slovaks and Hungarians from 1848 to 1945 and between the Czechs and Slovaks since the early 1920s.

Competition between rival Czech parties dates back to the early 1870s when two parties emerged to challenge the heretofore dominant National, or Old Czech, party — the Czechoslovak Social Democratic party, founded in 1878 and the National Liberal or Young Czech party founded in 1874. A true multi-party system did not emerge until the later 1890s with the formation of three progressive parties of the intelligentsia — the Radical Progressives, the State Rights Radicals, and Masaryk's Peoples, later Progressive, party — and three new mass parties — the Agrarians, the National Socialists, and the Christian Socialists. The most important underlying cause of their emergence was the growing social and occupational differentiation among Czechs wrought by industrialization and urbanization. More immediate causes of this emergence were the limited extension of suffrage in 1896 and the defeats sustained by the leading Czech political party, the Young Czechs, during the years 1897 to 1899. The belated (1896) Habsburg approval of the addition of a fifth, or universal, curia for elections to the lower house of the Parliament (*Reichsrat*) in the western half of the monarchy encouraged the formation of mass parties among all nationalities and marked the beginning of the end of the heretofore elitist parties of the notables. The failure of the predominantly middle class Young Czech party to implement very much of its program of extended civil liberties, universal male suffrage and greater use of the Czech language by governmental agencies and the courts also affected the formation of mass parties. The rural and lower middle class constituents of the party, already dissatisfied by its having often given precedence to commercial or industrial interests, now deserted it in large numbers to form the Czech Agrarian and Czech National Socialist parties. The long-established Czechoslovonic Social Democrats grew to become the leading Czech party in votes received, while the Czech Agrarians won more seats than any Czech party in the Bohemian diet as well as in the *Reichsrat*. The Czech clerical parties, the National Catholics and Christian Socialists — were encouraged by the encyclical *Rerum novarum* and won their greatest support in Moravia. The three progressive parties of the intelligentsia became two in 1908, attracted very little mass support, and continued to influence public opinion primarily through forthright and articulate journalism and the words and deeds of several outstanding individuals. The Czech parties established during the later nineteenth century and their Slovak counterparts became the leading parties of the first Czechoslovak Republic.

Another characteristic of Czechs and to a lesser degree Slovaks has been their ability to use political systems designed to favor other nationalities or classes to their own maximum benefit. This ability was especially pronounced among the Czechs in Austria-Hungary, where under constitutional rule from 1860 to 1914, Czechs enjoyed considerable powers of self-government at local and district levels as well as limited freedom to organize private or cooperative economic enterprises. Slovaks, in turn, took advantage of the many educational, cultural and political institutions established with Czech help in Slovakia during the 1920s.

The political tradition of the Czechs and Slovaks has been decidedly peaceful as well as pragmatic. Neither peoples have had a revolutionary tradition comparable to the neighboring Poles or Hungarians or to the relatively more distant French and Americans. Nonetheless, each people has tenaciously maintained its national identity if not its political independence. The Slovaks are very proud of their national uprising against the Nazis in the summer of 1944, undertaken at great peril and without the encouragement of the Czechoslovak government in exile or any Allied power. Although Czechs as well as Slovaks rose in 1848, they both acted circumspectly given their political inexperience and weakness, especially in contrast to the Hungarians. Great popular upheavals and changes of government in the recent Czech past have occurred with little or no bloodshed. Czech and Slovak leaders moved decisively and with widespread popular support in October 1918 to establish an independent Czechoslovak Republic, whose groundwork and Allied recognition as a co-belligerent had been largely established by T. G. Masaryk and others abroad. Against heavy odds and without allies, Czechoslovakia chose not to resist foreign invasion in 1938 and again in 1968. Czechs and Slovaks have nonetheless proved to be tough and resilient in most struggles with political adversaries, despite the fact that each people has in modern times usually preferred passive resistance to violence.

Given their rich heritage of the past in all areas of human endeavor, Czechs and Slovaks have endured severe political setbacks with greater equanimity than might otherwise be expected. That heritage not only inspires pride in past achievements but courage in present circumstances and continued confidence in the national future.

NOTES

1. An abridged version of this chapter was presented as a paper at the twentieth annual Missouri Valley History Conference in Omaha on March 11, 1977. I wish to thank the following colleagues who read the first complete draft of this chapter and made suggestions for its improvement: Professor Nicholas Bariss of the University of Nebraska at Omaha, Professor Carole Fink of the State University of New York at Binghamton, and Joseph Svoboda, University Archivist at the University of Nebraska–Lincoln.

2. The name Slovak (*Slovák*) like Slav (*Slovan*) derives from *slovo,* the word in all Slavic languages for "word" or "speech." A Slav (*Slovan*) is therefore someone who speaks the word (*slovo*) as opposed to the German (*Němec*) who is dumb (*němý*) in the sense of being unable to talk.

3. In very few cases did Czechs or Slovaks ever allow their enthusiasm for Slavic reciprocity to become enthusiasm for Pan-Slavism, that bugbear of German, Hungarian and Austrian officials and journalists from 1848 to 1918.

4. Anton Bernolak (1764–1813) had first proposed basing written Slovak on the Western dialect.

5. From the word *robota,* the term used to describe the forced labor owed by a peasant to a lord before the emancipation of the serfs of the Austrian Empire in 1848.

6. One should also refer to the works of Franz Kafka (1883–1924), a Jew from Prague who published in German and whose influence in Czech and Slovak literature made itself profoundly felt during the 1960s.

7. The figure for the Czechs includes members of the Czechoslovak Church as well as of Protestant Churches. Slovak Protestants are almost exclusively Lutheran.

8. Most historians have considered the Hussites to be primarily a religious movement with strongly Czech national social revolutionary overtones. Nineteenth century Czech historians often overemphasized the extent to which the Hussite wars became a struggle between Czechs and Germans. Contemporary Marxist historians have emphasized the revolutionary aims, actions, and implications of the Hussite movement.

9. The Jewish population of interwar Czechoslovakia always exceeded 350,000, roughly 2.5 percent of the total population.

10. The Czechs Matyáš Lerch (1860–1922), František Studnička (1836–1903), and Edvard Weyr (1852–1903) later also made original contributions to mathematical analysis.

11. Germans noted for their scientific work in Bohemia and Moravia during the nineteenth century included Christian Doppler (1803–1854), Professor of Mathematics in Prague and in Štiavnice during the 1840s and discoverer of the "Doppler effect," and Johann Gregor Mendel (1822–1883) of Brno, who discovered the general laws of heredity in plants and animals in 1865. Sigmund Freud (1856–1939), a German-speaking Jew, was born and spent the first four years of his life in the predominantly Czech town of Pribor in northeast Moravia.

12. World renowned Czech drivers of that era included Čeněk Junek, Eliška Janková, Jindřich Knapp, Jiří Lobkowicz, Zdeněk Pohl, Ella Slavíková, and Bohumíl Turek. Sources on the automotive industry and on racing are all in Czech.

BIBLIOGRAPHY OF SELECTED WORKS ON THE CZECHOSLOVAK HERITAGE UP TO 1918

Brock, Peter. *The Political and Social Doctrines of the Unity of Czech Brethren in the Fifteenth and Early Sixteenth Centuries* (The Hague, 1957).

Brock, Peter and Skilling, H. Gordon, eds. *The Czech Renaissance of the Nineteenth Century* (Toronto, 1970).

Dostal, O., J. Pechar, and V. Prochazka. *Modern Architecture in Czechoslovakia* (Prague, 1967).

French, Alfred. *The Poets of Prague: Czech Poetry Between the Wars* (London, 1969).

Harkins, William. *Karel Čapek* (New York, 1962).

Heymann, Frederick G. *George of Bohemia: King of Heretics* (Princeton, 1965).

Heymann, Frederick G. *John Žižka and the Hussite Revolution* (Princeton, 1955).

Kaminsky, Howard. *A History of the Hussite Revolution* (Berkeley, 1967).

Kerner, Robert Joseph. *Bohemia in the Eighteenth Century* (New York, 1932).

Masaryk, Thomas Garrigue. *The Making of a State: Memories and Observations, 1914-1918* (New York, 1927).

Okložilík, Otakar. *The Hussite King: Bohemia in European Affairs, 1440-1471* (New Brunswick, New Jersey, 1965).

Rabinowicz, Oskar, et al. *The Jews of Czechoslovakia*, 2 vols. (Philadelphia, 1968).

Rechcigl, Miloslav, Jr. *The Czechoslovak Contribution to World Culture* (The Hague, 1964).

Seton-Watson, R. W. *A History of the Czechs and Slovaks* (London, 1943, reprinted Hamden, Connecticut, 1965).

Seton-Watson, R. W. *Slovakia Then and Now* (London, 1931).

Šíp, Ladislav. *An Outline of Czech and Slovak Music*. Part II: *Slovak Music* (Prague, 1960).

Souckova, Milada. *The Czech Romantics* (The Hague, 1958).

Souckova, Milada. *The Parnassian Jaroslav Vrchlicky* (The Hague, 1964).

Spinka, Matthew. *John Hus: A Biography* (Princeton, 1968).

Štěpánek, Vladimír and Bohumil Karásek. *An Outline of Czech and Slovak Music*. Part I: *Czech Music* (Prague, 1964).

Sturm, Rudolf. *Czechoslovakia: A Bibliographic Guide* (Washington, D.C., 1967).

Thomson, S. Harrison. *Czechoslovakia in European History* (Princeton, 1953).

Wellek, René. *Essays on Czech Literature* (The Hague, 1963).

JOHN HUS

By the early 15th century the Christian church in the West was in deep trouble. The Great Schism, following the move of the popes to the French city of Avignon, resulted in the election of two, then three popes, each claiming exclusive rights over the papal bureaucracy and control over the church. These abuses deeply troubled true believers. Simony, the sale of church offices to the highest bidder, produced bishops who were unworthy aristocrats or even children whose families drew the lucrative revenues of their offices.

Papal taxation derived large amounts of money from the countries of Europe. Through the system of annates — by which newly appointed prelates were obliged to pay the equivalent of one year's income to the papal treasury — the popes increasingly exploited and commercialized church appointments. The popes enjoyed huge incomes, they lived in luxury and pomp, and exercised their official powers in an arbitrary manner. For all this the population — especially the peasants and the inhabitants of cities and towns — paid heavily since all the funds were taken from them through various forms of tribute and taxation. And as if this were not bad enough, many unworthy priests and monks lived dissolute lives, mocked the rules of celibacy and sobriety, and provoked a growing contempt among the population for the priestly professions. A system of indulgences — church-promoted sales of pardons for lesser sinners, supposedly shortening their stay in purgatory after their death — further lowered the prestige of the church among the population. All this was happening at a time when national monarchies were emerging in many areas of Europe, challenging the political and economic power of the dominant universal institutions, the Christian church and the Holy Roman Empire.

Honest churchmen perceived their difficult task as consisting of two parts; first, the unity of the church had to be restored by deposing the rival popes. Secondly, a thoroughgoing reform of the personal lives and attitudes of the clergy in general had to be instituted. Above all, discipline in the ranks had to be reestablished as a step toward regaining the lost prestige of religious institutions.

The calling of a general council seemed best suited for the accomplishment of the first task. Precedents were not lacking for such a solution; previous councils dealt with a variety of troubles that beset the Christian church, including the definition of church doctrines. Now a council was to gather at the Italian town of Pisa in 1409. The council was determined to replace the quarreling popes by a newly elected head for the church. This assumed that the council would be superior to the pope in church affairs, an idea that was by no means universally accepted by all leaders of the religious institution.

The ruler of the Holy Roman Empire and the King of Bohemia was Wenceslas IV of Luxemburg; he was the brother of King Sigismund of Hungary. Wenceslas' rule was shaky in Germany. Sigismund was a stronger personality who often took advantage of Wenceslas' weaknesses. He supported the idea of the convocation of a church council because he aspired to the office of emperor and wanted to appear as the peacemaker of Western Christendom. This would gain him the support of the electors of the Empire.

In most European nations the scandalous state of the church caused great concern and anguish. Reformation was, therefore, a prime necessity. Rulers and pious churchmen alike desired it, laymen and simple priests hoped for it. But there was little agreement on the practical steps that had to be taken in the interest of reform, despite the willingness of the leaders to call for the convocation of a council.

In the Czech lands the living conditions of the people were not greatly different from those of the neighboring countries. Here, however, the problems caused by serfdom and the exploitation of peasants and city dwellers by a relatively small number of noblemen were exacerbated by conflict between different ethnic groups. This conflict began in the towns and cities. The Bohemian towns were heavily populated by German craftsmen and merchants. Their forefathers or they themselves were invited to Bohemia by various kings during the late Middle Ages. The skills of the craftsmen and merchants were needed in building up the cities and in developing trade and commerce. However, the Germans were given too many privileges not shared by the Czechs. This was especially evident in their control of the city and town administrations from which Czechs were usually excluded. By the early 15th century, the Czech population of the cities, especially that of

Prague, was increasing. They were resentful of German privileges and wanted privileges for themselves. This led to all sorts of conflict.

The establishment of the University of Prague in the mid-14th century provided a forum for many of these conflicts. At first, the control of the university's administration was in the hands of German masters, appointed by the King.

Wenceslas IV was a weak ruler. Relations between Wenceslas and the church in Bohemia were difficult. The situation was aggravated by the elevation of a strict, ascetic archbishop, John of Jenstein, to the See of Prague. The archbishop was a man of rigid disposition. But his strictness was misdirected; instead of using his considerable energies for the renewal of the Bohemian church, he was determined to preserve the considerable privileges of his institution against all comers, including the crown. Consequently, there were bad feelings between the archbishop and King Wenceslas. In 1393, Wenceslas decided to establish a new bishopric and endow it with the properties of the monastery of Kladruby. The archbishop was adamantly opposed to this plan. Wenceslas, who believed that the advisors of the archbishop were responsible for his obstinacy, ordered their imprisonment. One of the archbishop's favorite advisors, the vicar general John of Nepomuk, was tortured and eventually died.

Many Czech aristocrats resented the high-handed manner in which the King treated the clergy, and feared that their own privileges would be questioned next. They banded together in a league to demand that Wenceslas dismiss his advisors. They wanted him to appoint only aristocrats to state offices. When Wenceslas refused, he was imprisoned by the leaders of the league. At this time Wenceslas's younger brother, Jan, Duke of Zhorelec, saved him; the royal towns also helped the King by mobilizing their forces and threatening the higher nobility. Finally a compromise was achieved through the help of King Sigismund of Hungary, by which Wenceslas was to share some of his power with the Czech aristocrats.

In the meantime, Wenceslas changed the orientation of his foreign policies. Originally, he supported the Roman pope against the one in Avignon, Boniface IX against Clement VII. In line with this policy he sought an alliance with Richard II, the King of England, the enemy of the French. This alliance was concluded by the marriage of Richard II to the sister of the Czech King, Anna. But soon Wenceslas changed course. He agreed with Charles VI of France that both the Roman and the Avignonese popes should be deposed by a council of church fathers and a new pope elected in their place. Naturally, his former ally,

Boniface IX, did not take kindly to this change in policy. He encouraged the German electors to depose the Czech King as Emperor in 1400, and they elected Rupert of the Palatine in his stead.

The Bohemian aristocrats felt encouraged by this decision and renewed their struggle for supremacy in the Czech state. In order to counter their efforts, Wenceslas invited his brother Sigismund to Prague to rule with him as co-regent, hoping that he would act as an intermediary in his attempted reconciliation with the Roman pope. But Sigismund betrayed his brother and took him prisoner. In turn, the towns in Bohemia revolted against the Hungarian King and helped Wenceslas to escape. Finally, Wenceslas concluded an agreement with his Czech opponents in 1403, but this did not end the struggle. It was further worsened by a revolt of the clergy of Bohemia against the policies of their King.

The church in Bohemia was exceptionally rich by the early 15th century. It controlled almost half of the arable land of the country, and its monasteries were well-endowed with worldly goods. Most Czechs regarded an ecclesiastic office not as a means to support their spiritual betterment, but as a secure job ensuring its holder a better-than-average existence. Since the King was the supreme patron of the church, he often used church offices to reward his followers with benefices. In turn, the wealth of the church lessened the desire of many of the clergy to live according to the rules of the Christian religion. Some of them became very wealthy and enjoyed luxuries not available to the common people. Many of the rich priests hired substitutes to celebrate the mass for their charges while they spent their time and money abroad or at the royal court or in the company of noblemen. On the other hand, poor priests who were not well connected and usually came from the lower ranks of society, lived difficult lives. They were often critical of the luxurious life of their fellow priests. They also resented the large sums of money that were collected in Bohemia through papal taxation.

Several of these priests raised their voices against the excesses of the Bohemian clergy. Konrad of Waldhauser, and Milic of Kromeriz were among these, and they soon gained many followers among the lower clergy. Milic maintained that the antichrist was about to descend on earth because of the general corruption, immorality and love of luxury that dominated the life of the clergy. He was eventually burnt at the stake at Avignon where he had been summoned for an examination of his views. But the disciples of these men were numerous and they continued their work against clerical abuses in Bohemia.

The activities of these reformers found an eager reception among the population. Their movement grew, especially after the foundation of the Bethlehem chapel in Prague by a rich merchant called Kriz, who wanted a place for Czech-speaking priests to preach to the people. This was the chapel that was to provide a forum for a young priest, John of Husinecz.

John of Husinecz is better known by his abbreviated name of John Hus. He was born in 1371. He studied at the University of Prague and received the Master of Arts degree in 1396. In 1398, he became a lecturer at the school and two years later was ordained as a priest. In 1401, he was elected dean of the faculty of philosophy. By 1402, Hus had developed into an effective orator and was appointed to the Bethlehem chapel. He was universally respected by his colleagues for living an exemplary life. He was totally dedicated to his priestly calling and soon became involved with the reformers.

Hus was extraordinarily successful as a preacher at the chapel. His sermons attracted large audiences from all segments of the population, including the clergy. The archbishop Zbynek of Hasenburg encouraged Hus in his efforts to root out the abuses from his administration and twice invited him to speak at synods of the Bohemian church in 1405 and 1407.

Some clergymen, however, were offended by the activities of Hus and they objected to his sermons. These priests accused Hus of exaggerating clerical misbehavior for his own purposes and degrading priestly dignity before the population. But this did not deter Hus from his course; he continued preaching against the abuses in the church and began to evaluate these abuses in light of church doctrines. He was deeply impressed by the teachings of the Englishman John Wycliff, whose works had reached Prague through Czech students who studied at Oxford.

Wycliff rejected nominalism. But it was not his philosophical treatises that attracted Hus but the Englishman's views on the values of church traditions. Wycliff was a true reformer. He demanded that the clergy accept the essence of the teaching of Jesus which, for him, meant the rejection of worldly riches as a hindrance to the salvation of the soul. He stated that the wealth accumulated by the church was the dowry of Satan and as such should be confiscated, preferably by the state. Wycliff considered the church to be an invisible society composed of individuals who were predestined by God for salvation. This was a new version of the ancient Donatist notion, condemned by the council of Nicea in 325, that the church was an assembly of saints from which all sinners were excluded by God's order. Naturally, such a church, headed by Jesus himself, did not need the hierarchy or the clergy, and it would especially reject the papacy as its major institution. Wycliff maintained that the office of the pope was actually invented by the Antichrist. He also rejected the veneration of saints and relics, the institutions of celibacy and monasticism, since these were, according to him, hindrances to a good Christian life and salvation.

Jerome of Prague, a young friend of Hus, may have been instrumental in spreading Wycliff's theology in Prague. The notions of the English reformer were read and soon endorsed by many members of the university's faculty. Hus himself quoted long passages from Wycliff in his sermons, but he did not simply copy the ideas of Wycliff. Hus was attracted more by the call to action voiced by the reformer and accepted only those of his ideas that were compatible with his own thinking. But Hus could not go on for long before being challenged by his opponents at the university. The German professors were especially opposed to his teachings and they proceeded to condemn the Wycliff doctrines in 1403, a resolution that was not fully endorsed by the Czech members of the faculty. When the archbishop sided with the Germans, Hus protested against the condemnation of preachers who supported Wycliff's notions and he lost the confidence of his ecclesiastical superior.

The situation was soon worsened by international developments. The Council of Pisa opened in 1409. In order to win the support of King Wenceslas, the council promised him that it would support his claim to the emperorship if he abandoned the Roman pope and accepted a new head of the church about to be elected by the council. Wenceslas agreed to this proposal. But the German masters at Prague University rose up against him.

Wenceslas needed the support of the university for the fulfillment of his plans in the Empire. Thus, he proceeded to alter the university's constitution, giving the Czech professors the majority of the votes. At this, the German masters and students left Prague University and went to various German schools in the Empire. They also spread rumors about the alleged triumph of heretics and heresy in Prague. Finally, the university voted to support the King's preference for the papal office. Now Wenceslas was free to recognize Alexander VI, elected to the papal office by the cardinals at Pisa. But the archbishop of Prague vehemently opposed the King's choice, and relations between the two deteriorated further.

John Hus and his supporters accepted Wenceslas' choice. But soon after the vote at the university they found themselves embroiled in new controversies. Although Hus was elected rector of the university after the departure of the Germans, the archbishop denounced him at the papal curia at Rome as a dangerous aberrant. The pope himself ordered that the Wycliff teachings must stop at Prague. Thereupon the archbishop declared a moratorium on preaching outside parish churches, a decree that was obviously directed against the preachers of the Bethlehem chapel. But Hus disregarded this order and appealed to the new pope, John XXIII, who succeeded the late Alexander VI.

The archbishop now ordered the condemnation of all the teachings of Wycliff, and he had all the Englishman's manuscripts found in Prague openly burnt in the courtyard of the palace. He followed this up by excommunicating Hus and his friends for disobedience. Hus appealed to the pope once again and continued to preach and teach at the university. His excommunication was confirmed by Rome in March, 1411, and when he continued to disregard the verdict, the archbishop placed the city of Prague under interdict. This stopped the performance of all church services, including marriages, baptisms and funerals. When Wenceslas decided to use force to break the ban, the archbishop attempted to flee to Hungary but died on the way.

A new incident increased tensions even more. This was the excommunication of Ladislas of Naples by Pope John XXIII, since the former refused to recognize the pope and started a war against him. In order to pay for the expected campaign, the pope offered indulgences to those who donated funds for his army. The papal bull offering the indulgences was sent to every court in Europe. When it was read in Prague, Hus immediately spoke out against it. He decried the "selling" of these pardons to the population by "unworthy priests." Hus also began preaching about the questionable doctrinal bases of the granting of the pardons. But there was a divergence of opinions between Hus and his friends at the university and many debated with him over this issue. Soon riots broke out in Prague interfering with the distribution of the indulgences. The King still supported the Roman pope; and he now threatened everybody who interfered with the indulgences with death. Three young men were actually executed when they disobeyed the royal orders. At this, the population of Prague rioted anew. The bodies of the executed men were brought to the Bethlehem chapel where John Hus buried them. He also celebrated a funeral mass in which he called the young men holy martyrs. The reformer thus angered his sovereign and lost his support. The King ordered Hus to leave Prague and he also sent his opponents into exile.

During the summer of 1412, Hus was once again excommunicated, now on account of his refusal to appear in Rome for an examination of his views. He retired to a castle in southern Bohemia, where he wrote several books, some of them in the Czech language. He also preached several sermons to the peasants who went to hear him in great numbers. Occasionally, Hus visited Prague and continued preaching at the Bethlehem chapel.

In 1414, the Council of Constance was convened to finish the job of the Pisa council. By then, King Sigismund of Hungary was the Holy Roman Emperor, Rupert having died in 1410. Sigismund was most anxious to settle the Bohemian troubles since he aspired to rule the Czech lands. As a first step, he offered safe conduct to John Hus to come to the council and present his ideas to the assembled fathers. Hus accepted this invitation. He trusted Sigismund and expected to receive a fair hearing at the council. The problem was, however, that the council was more concerned with the restoration of the unity of the papal office than with the reform of abuses. Since deep disagreements over the possible solution to this problem continued to divide the participants, the case of Hus came as a heaven-sent opportunity, as a welcome diversion from a seemingly insoluble dispute.

Hus traveled through Germany in the company of a few of his friends. His ability as an orator gained him respect among the population of the towns in which he stopped. In the city of Nürnberg, he was cordially received and accorded great honors after he delivered his sermon in the church. His expectations for a fair hearing and an honored place at the table of the church fathers seemed to have been justified. People everywhere crowded the churches where he spoke and uniformly praised his apparent honesty and the brilliance of his reformist zeal. But Hus committed a fatal mistake. Instead of going to the town of Speyer where the emperor Sigismund was staying in preparation for his trip to Constance, Hus went directly to the meeting place of the council. Thus, instead of arriving with the train of the emperor — after having possibly convinced the sovereign of the validity of his ideas — Hus arrived in Constance in the company of some relatively insignificant men. When he arrived, he continued preaching despite the fact that the ban of excommunication was not lifted from him. Not only did he preach, but he also celebrated the mass, and the population eagerly listened to his sermons.

Hus' accusers included John Zelezny, the bishop of Litomysl,

Michael de Causis and Stephen Pálec, all three of them Czech Catholics. They charged that Hus promoted heretical notions and disobeyed the orders of the church by his defiance of the ban. At their instigation Hus was locked up in the local Dominican monastery. But he was not badly treated as yet; he was able to write letters and communicate with his friends as well as prepare for his defense at the coming examination. He also hoped that when the emperor arrived he would be freed in observance of the promised safe conduct.

Sigismund arrived at Constance on Christmas Eve, 1414, and was dismayed to find that his guarantee of safety was disregarded by the fathers. But he soon found that this was simply the result of the desire of the cardinals and bishops assembled in council to show that they were independent not only of the papal office but also of the control of the emperor. After some deliberations, Sigismund decided to sacrifice Hus to his own interests. By accepting the treatment accorded to the Czech reformer as justified, Sigismund came to an agreement with the fathers and gained their support for his plans in the Empire. His spokesmen then asserted that the emperor's safe conduct was valid only to ensure Hus' safe travel to Constance, but it did not promise immunity in case he was found to be a heretic. This flimsy excuse sealed the fate of John Hus.

Hus was placed under the supervision of the bishop of Constance and was thrown in the Gottlieben castle jail. He was kept in chains. When, upon repeated examination, he refused to recant his views and called on the examining churchmen to prove him wrong by scripture, he was thrust into the deepest dungeon and was deprived of food and water. His clothes rotted on him; he lost most of his teeth to malnutrition and was barely alive when he was brought forth for his final examination.

A special commission proceeded with this examination. First, the doctrines of Wycliff were condemned once again and declared heretical. Hus was given three chances to recant. Instead, he tried to prove the correctness of his beliefs by quoting scripture as the sole authority for Christian doctrines.

The examining magistrates accused Hus on six counts. The first of these was that he did not believe the doctrine of transubstantiation, a notion that maintained that the wafer and the wine used by priests during the celebration of the mass was literally transformed by his actions into the body and blood of Christ. Hus argued that the notion of transubstantiation was merely a symbol, and that it was not proclaimed by Jesus, but by canon law, after six hundred years had passed.

Then he was accused of rejecting the doctrine of the infallibility of the pope in matters of faith, the veneration of saints and of their relics. In reply, Hus challenged the fathers to show him where such a doctrine could be found in the Bible. He further argued that the pope was human and had faults just like other human beings, and that scripture actually stated that "nobody is perfect, only God alone!" Next he was accused of rejecting absolution by priests who were themselves sinners and having excused people from making confessions. Hus admitted that this was true, reflecting the ancient Donatist teaching that the church was to be composed of saints, and sinners had no place in it. Hus was also accused of disobedience and of rejecting the rule of celibacy for priests. He answered that scripture ordered obedience only to God, and that it was sinful to obey senseless laws. The canon law ordering priests to be celibate, he argued, was just such a senseless law. He also raised the question of indulgences, arguing that these served no other purpose than to maintain the popes in a state of luxury, and that this was a sin against the teachings of Jesus.

After this, the end came swiftly for John Hus. He was declared a heretic by the fathers; 51 votes were cast against him and 37 on his behalf. Then 45 voted for his death at the stake, 11 for public excommunication and sparing his life, and 30 voted not guilty. These votes did not go along ethnic lines; many German fathers sympathized with the reformist ideas of Hus and voted for his release. On the other hand, some Czechs sided with his enemies.

Finally, Hus was asked for the last time to recant his views in order to save his life. He refused. Therefore, on July 6, 1415, he was burned at the stake. As a chronicler noted, he died of smoke before the flames touched his body. A few months later the council decided that the close friend and collaborator of Hus, Jerome of Prague, who arrived at Constance during the condemnation proceedings and stood by his friend during the trial, was also a heretic. When he refused to denounce Hus' teachings, Jerome, too, was burned at the stake. Thus the movement of Hussitism was born out of the smoke of Constance.

John Hus was a martyr for his beliefs. He was also a champion of individual liberty and a forerunner of the mighty tide of reformation that was to sweep Europe in the 16th century. But his failure to effect a reformation of the ways of the church resulted more from political than spiritual factors. It was obvious to everyone that, by that time, the church would have to be changed if it wanted to survive. But reform could no longer be brought about without active support from the political authorities. Hus, like other clergymen seeking reform who could not inspire confidence in the new national monarchies or princes, faced certain death at the stake. This was a lesson that Martin Luther was to learn a hundred years later.

After the trial at Constance, the Czechs were up in arms against both the council and the emperor for permitting this outrage to happen. Czech national consciousness was strengthened by the trial and execution of Hus. The Bohemian diet immediately passed a resolution to refuse obedience to the pope unless his decrees could be proven valid by scripture. The diet also agreed that, from then on, Prague University would be the sole judge of heretical doctrines in Bohemia, and that no Czech would be permitted to be tried by outside authorities.

But the Hussites were not speaking with one voice. Deep divisions appeared among the faithful, mainly along social lines. Among the peasants and the burghers the doctrines of the martyr became the foundations on which a mighty social movement was to be built.

Just before he left for Constance, Hus approved the serving of the communion to laymen in two kinds, in line with his teaching that the eucharist was merely a symbol of the sacrifice of Jesus. Most of the moderates who followed Hus' teachings were called Utraquists. They were named after the chalice in which they served the wine to laymen during the celebration of the mass. The moderates wanted to restrict reform to church practices, arguing that the changing of the ways of this world was not their task. Many Czech noblemen belonged to this group. But there were also radical followers of Hus in Bohemia. They went beyond mere theological issues and began to teach that the aim of the Christian religion should be to bring social justice to mankind. The radicals established a camp which they called Tabor, and they began building an army composed mainly of the poor from the countryside. Thousands of peasants flocked to this camp showing that the people were attracted by the radical ideas.

The radicals and moderates agreed on four principles as the essence of Hussitism:

1. preaching by anyone who believed in God and who did not have the holy orders conferred on him was permissible and even desirable;
2. the communion was to be served to all people in two kinds;
3. all worldly possessions of the secular and regular clergy should be confiscated, since their wealth was a sin against God;
4. simony and other deadly sins were to be severely punished, sometimes by death.

The radicals added many other "theses" to these four, although they all tended to be variations on the theme of social equality. Both factions were suspicious of the Luxemburg rulers, and were determined to resist the willingness of these rulers to suppress Hussitism.

King Wenceslas, as if he wanted to prove the suspicion of his subjects valid, placed an anti-Hussite town council in power in Prague in 1419. Thereupon the radicals revolted and threw the new councillors out the windows of the town hall. However, within a month, Wenceslas was dead and King Sigismund declared himself the inheritor of the Bohemian crown.

The Hussites, naturally, regarded Sigismund as their arch enemy, the man who deceived Hus and brought about his death. No wonder then that there were few Czechs to support Sigismund's claim to rule. The dependencies of Silesia, Lusatia and Moravia were, however, ready to accept him as their ruler since these provinces contained large German populations and the Germans found nothing wrong in Sigismund's actions. But the Czechs of Bohemia would accept Sigismund only if he agreed to the Articles of Prague — and Sigismund refused. Instead, he proclaimed a crusade against the Hussites. Thus began a long series of wars which were to devastate large areas of Central and Eastern Europe and in which the Hussite warriors were usually victorious. They proved, incidentally, that the era of the knightly armies of the Middle Ages had ended; their modern, innovative techniques based on the use of infantry and mobile forces invariably defeated the cumbersome, armored knights. Their methods of warfare proved to be the harbinger of the tactics of popular armies of later times.

Two large armies were led into Bohemia by Sigismund in 1420 and 1421. Both were soundly beaten by the Hussite army led by a one-eyed general, Jan Žižka. This was an army manned mostly by the radical wing of the Hussite movement. Žižka himself was a veteran of many wars and he introduced new tactical concepts. These were based on fortifications composed of heavy wagons, chained together, which served to defend the peasant warriors against the cavalry charge of the armored knights. When the knights exhausted themselves in futile attacks on the wagon-fortresses, the chains were opened and the wagons were used as attack vehicles. The peasants used these wagons with great skill, maneuvering to cut off parts of the cavalry of the attackers and destroying them piecemeal. Žižka also effectively used cannon on his wagons as field artillery. These armies were terribly swift in attack and were universally feared by their enemies. Other generals emerged from the army of the radical Hussites, among them the two Prokops and, perhaps the most famous of them, Jan Yiskra (Giskra).

Despite the fact that he succeeded in having himself crowned in Prague in 1420, the Hussites did not recognize Sigismund as their king. Thus, they invited the Polish prince, Sigismund Korybut, to come to Prague as their ruler. But Korybut also tried to effect reconciliation with the Roman pope and he, too, was eventually ousted. Now the Bohemian diet — in which the towns, especially Prague and Tabor, were strongly represented — exercised power in Bohemia. During this time simmering divisions came to the fore among the Hussites themselves.

Prague was the largest Bohemian city, the capital of the kingdom. Its population was close to 20,000 inhabitants. It was a major trading and commercial center, drawing to itself a large part of the eastern trade. It was also a major intellectual center and the University of Prague played a major role in European culture and politics. Most professors were moderate Utraquists who were quite satisfied with the religious rules established in the Articles of Prague and who wanted little to do with the egalitarian ideas of the radical Taborites.

Tabor, on the other hand, stood for radical changes in society. Its population included several foreign heretics who were driven out of their country and congregated at this place. They accepted the chiliasm and millenarianism typical of radicals in this century. The English Lollard, Peter Peyne; former serfs who were declared free at Tabor; poor priests who joined the people against their higher clergy, all went to Tabor. They believed that the second coming of Christ was near and were preparing for the final war of extermination against the enemies of "true" — Hussite — Christianity. They rejected canon laws and the traditions of the church as sinful. They proceeded to destroy the symbols of these traditions, the altars and decorations in the churches, wherever they went and they looted and destroyed the monasteries. They elected their own priests and rejected any notions of a church hierarchy. They wanted nothing to do with the established church which they declared to be the vehicle of the Antichrist, and the idea of reconciliation with Rome was hateful.

The Taborites were not only religious, but also political and social radicals. At Tabor, serfdom was declared abolished; in its place, a commune of the people was proclaimed in which all would share equally in the available goods. Another radical group formed the Camp of the Orphans in Eastern Bohemia. This camp consisted of soldiers who served in the Army of Žižka who died in 1424, and regarded themselves the orphaned children of the great general. It was inevitable that the divided house of the Hussites would face a serious challenge not only from without but also from within. As long as the threat of a crusade hung over them, all factions united against the common enemy. But when the danger of invasion lessened, the Hussites began fighting among themselves. Irreconcilable social differences, therefore, played into the hands of their enemies.

The Utraquists perceived the germs of social disorganization in the ideas prevalent at Tabor and were fearful of their consequences. They began to look for ways through which reconciliation with the Catholic minority in Bohemia and, in turn, with the papacy, could be arranged. Although great victories over their foreign enemies in 1427 and 1431 enabled them to hold their own against the church, and the church council of Basel, called together in 1431, invited their representatives once more to discuss the possibility of a compromise, Hussite reformism was actually receding. The Czech delegation to the council — whose representatives met the Czechs at Eger in 1432 — was led by Jan Rokyczana, the vicar of the archbishop of Prague. Although no agreement was reached at this time, the negotiations continued in Prague. The representatives of the council finally convinced the Utraquists to accept a so-called Compact, by which the Articles of Prague were rendered harmless by stating that of all the Hussite reforms, only the chalice, conceded to laymen, should remain in effect. No new Hussite priests were to be created, although priests who joined the movement would be permitted to serve in office until their death.

While the negotiations were going on in Prague, the radicals and the Utraquists finally came to blows. The nobility and the Czech Catholics sided with the moderates and they fought a decisive battle at Lipany on May 30, 1434. Prokop the Bald, one of the great Taborite generals, was killed; the radicals were overthrown and driven out of Bohemia. Soon an agreement with Sigismund stipulated that he was to be recognized as King of Bohemia in exchange for his promise to tolerate the existing practices of the moderate Hussites. The election of bishops was to be given to the Bohemian diet, and Rokyczana was to be elected archbishop of the Czech lands. The Compact of Prague which the delegates of the council of Basel negotiated with the Utraquists was accepted. The struggle of the Hussites and Catholics thereupon entered another phase.

The consequences of Hussitism in Bohemia were great indeed. First, they led to the confiscation of vast church estates and other properties which were now in the hands of the Czech nobility, especially the aristocracy. Thus, the high nobles were able to act as equals to the King, and an entirely new situation was created in the relations between the crown and the noble orders. The latter were soon calling themselves the political "nation" to the exclusion of other segments of society. On the other hand, the lower classes of society, especially the peasants, lost most of what they had initially gained. Large numbers of them were killed in the Taborite wars, hundreds of villages were destroyed, their population dispersed. Many of the displaced peasant warriors provided a constant source for banditry and other forms of lawlessness, becoming a menace to the neighboring peoples as well as their fellow Czechs for several decades. In general, Bohemia as a whole was impoverished and its population was greatly reduced.

Nor were the consequences of the Hussite wars less severe for the neighboring peoples, especially for the Germans and Hungarians. Even before the destruction of Tabor, radical Hussite armies regularly entered the German and Hungarian lands in retaliation for the devastations caused by the German and Hungarian troops in Bohemia. They raided and destroyed villages and looted towns and monasteries. Interestingly, these armies often found sympathy among the poor peasants in the countryside who regarded the Hussites as their liberators from their landlords.

However, Hussitism also left an important intellectual legacy whose impact was equally significant, especially in Hungary. Several students from southern Hungary studied at the University of Prague during the late 14th and early 15th century. These students learned Hussite doctrines during their stay in Prague. Jerome of Prague himself preached at Buda, the royal residence, in 1410 and was expelled by King Sigismund. By the 1430s, Hussite teachings were well known in Hungary, especially in the south.

It seems that peasants were especially interested in Hussitism in Hungary, and the Taborite radical version found many followers. There were soon reports of secret meetings of peasants in abandoned mills or in forests, where entire villages were instructed by fugitive Hussite missionaries about social, religious and political equality. These reports greatly alarmed the Hungarian clergy, and their remonstrances against the Hussites found willing listeners at the court of King Sigismund. The consequence was the invitation of a Franciscan monk, Jacob of Marchia, to Hungary to act as papal inquisitor and to root out the Hussites.

Jacob of Marchia was well known for his ruthlessness in fighting heresy in the Balkans. In 1434 he went to Hungary and conducted a war of extermination against the Hungarian Hussites. The number of his victims is not known. However, he used the death penalty freely, and he even ordered the disinterment of the bones of suspected heretics who perished before he could punish them. All in all, Jacob asserted that he "saved the souls" of 25,000 persons in Hungary by reconverting them to Catholicism. If this number seems exaggerated, it certainly indicates the success of Hussitism in the Hungarian countryside.

Several Hungarian clergymen escaped the persecutions together with their flocks and they went to Moldavia where they found a new home. Here they translated the Bible into Hungarian. The language of this Bible became the foundation for the literary language of the Hungarians.

Finally, a great peasant revolt broke out in Transylvania in 1437. Its leaders indicated in their demands that they, too, had been influenced by the Hussite teachings. Although this revolt was ruthlessly suppressed, it indicated the fact that Hussitism broke out of the confines of the Czech language-area, and its doctrines became an inspiration to people throughout Eastern Europe.

In 1437, King Sigismund died. His successor, the Habsburg Albert II, reigned for only 9 months. Upon his death, George of Podiebrady, a nobleman from Eastern Bohemia, was elected the governor of the Czech lands for two years. When the posthumous son of Albert II, Ladislas V, grew into manhood in 1452, he was elected King of the Bohemian lands under the guidance of Podiebrady. But the young king died in 1457. The next year George of Podiebrady was elected King of Bohemia by the diet, and with him a moderate Hussite came to power. With him a new age began in the history of the Czech lands.

BIBLIOGRAPHY

NOTE: This bibliography provides only samples of the enormous literature that exists on John Hus and Hussitism. Only English language sources are included.

Kaminsky, Howard. *A History of the Hussite Revolution* (Berkeley, California: University of California Press, 1967).
Kitts, Eustace, and John. *Pope John the Twenty-Third and Master John Hus of Bohemia* (London: Constable, 1910).
Macek, Jozef. *The Hussite Movement in Bohemia* (Prague: Akademia, 1958).

Okloźilík, Otakar. *Wycliff and Bohemia* (Prague: Akademia, 1943).

Poggio-Braccolini. *The Trial of Hus, His Sentence and Death at the Stake in Two Letters to a Friend, Written by a Participant of the Council of Constance* (New York: Granville, 1930).

Spinka, Matthew, trans. *The Letters of John Hus* (Manchester: Manchester University Press, 1972).

_____. *John Hus. A Biography* (Princeton, N.J.: Princeton University Press, 1968).

_____. *John Hus' Concept of the Church* (Princeton, N.J.: Princeton University Press, 1966).

Victor S. Mamatey

THE BIRTH OF CZECHOSLOVAKIA:
UNION OF TWO PEOPLES

When the Austro-Hungarian government declared war on Serbia on July 28, 1914, which precipitated the First World War, the Czech and Slovak people were filled with a foreboding that the great conflict would deeply affect their destinies. No one had a clear vision of the future, however. At the beginning of the war, there was no Czech or Slovak revolutionary movement aiming at the break-up of the Habsburg monarchy and the establishment of a Czechoslovak state.

The Czechs of Bohemia

The Czechs and Slovaks had been a part of the Habsburg empire for four centuries. In 1526, the feudal estates of the kingdom of Bohemia and the feudal estates of the Austrian duchies and the kingdom of Hungary formed a personal union under the Habsburg dynasty. Within a century, however, Bohemia's voluntary association with Austria and Hungary was transformed into an involuntary union under an alien, hereditary, absolutist dynasty. In 1618, outraged by violations of their political and religious rights, the Protestant estates of Bohemia rebelled against the Habsburgs. Two years later, their army suffered defeat in the fateful battle on the White Mountain, which decided the fate of the Czech people for the next three centuries. Bohemia's sovereignty and independence were extinguished and the country was absorbed into a common Austro-Bohemian state, dominated by the Germans. Even before the Battle on the White Mountain, Bohemia had a large German minority. Now the Czechs lost their Protestant nobility and, to a large extent, also their class of townsmen. They were reduced to a nation of peasants, until the nineteenth century when they effected a remarkable economic, social, and cultural revival.

In the nineteenth century, the lands of the Bohemian crown (Bohemia, Moravia, and Silesia) were swept by the Industrial Revolution, making them the industrial center of the Habsburg empire. About 80 percent of the empire's industries were concentrated there. Industrialization naturally wrought deep changes in Bohemian society. Out of the Czech peasantry, which itself underwent a remarkable evolution, there grew a politically well-organized, nationally conscious Czech industrial working class and a prosperous, well-educated and intensely nationalistic Czech middle class. Conscious of their common peasant origins, all classes of the Czech people showed a remarkable national solidarity. The Czech national ethos was intensely nationalistic and profoundly democratic and egalitarian.

Guided by the motto "in work and knowledge is our salvation" ("*v práci a vědění jest naše spasení*"), the Czechs built up, during the course of the nineteenth century, their national and cultural institutions until these became the envy of the other peoples of the Habsburg empire. The Czech National Museum, founded in 1818, gathered and exhibited the memorabilia of Bohemia's colorful medieval history, instilling in the Czechs a pride in their past. The organization of the Czech *Matice* in 1831, on the model of the Serbian *Matice,* provided a clearing house for Czech national literature. The gymnastic organization "Falcon" ("*Sokol*"), established in 1862 to promote the idea of "a sound mind in a sound body" ("*mens sana in corpore sano*"), proved to be perhaps the most effective instrument for the promotion of Czech national pride and solidarity; Czechs of all classes met in it as equals (members addressed each other as "brother"), held together by the love of their nation and country.

The division of the University of Prague in 1882 into a Czech university and a German one emancipated Czech scholarship from German domination and provided a training ground for a distinctly Czech national intelligentsia. The festive opening of the handsome new building of the Czech National Theater in Prague in 1883, the construction of which was made possible by public subscription, provided a dignified setting for the staging of Czech plays and operas, especially the operas of the great Czech national composer Bedřich Smetana (1824–1884). The establishment of the Czech Academy of Arts and Sciences (1890) and the Czech Philharmonic Orchestra (1901) rounded out the organs of Czech national culture. Although inspired by nationalism, the Czech cultural institutions were not narrowly provincial in character. They were patterned on the best models in Europe and achieved high standards of professional excellence.

Czech Political Aims

By 1914, the Czechs had attained a level of social, economic and cultural development second only to that of the Germans in the Habsburg empire. They had failed to achieve political equality, however. The aim of Czech nationalists was to restore Bohemia's historic "state right," — to restore the sovereignty of the kingdom of Bohemia and secure for it a place in the empire analogous to that of the kingdom of Hungary after the Austro-Hungarian Compromise of 1867. Emperor Francis Joseph I (1848–1916), perhaps, might have conceded the Bohemian state-rights program. Three times he promised to have himself crowned with the crown of St. Wenceslas and grant the Czech demands. Each time, however, the Germans of Bohemia (called Sudeten Germans after 1918), who feared isolation in an autonomous Bohemian state with a Czech majority, managed to frustrate an agreement between the emperor and the Czechs by mobilizing German nationalist opinion in other parts of the empire against it.

The Bohemian Germans and the Czechs reached an impasse, which tended to drive them into the Pan-German and Pan-Slavic movements, respectively. Apprehensive lest the emperor make a deal with the Czechs at their expense, many Bohemian Germans began to look across the Austro-Hungarian frontier towards Imperial Germany and call for their return "home into the empire" (*"heims in Reich"*). They reproached Bismarck for failing to "complete" German unification, and insisted that Austria and Bohemia as lands of the former Holy Roman Empire (962–1806) and the German Confederation (1815–1866) must also be included in Germany.

The Pan-German agitation frightened the Czechs. While growing increasingly disaffected in the decaying Habsburg empire, they had no wish to be incorporated into the powerful new German empire. In these circumstances, many Czechs began to look toward Russia for salvation. Foremost among them was Karel Kramář (1860–1937), the leader of the nationalist Young Czech party, which stood on the right of the Czech political spectrum. Shortly before World War I, he proposed to the Russian government the formation of a great "Slavic imperium" under the Romanov dynasty in which the Bohemian kingdom would constitute an autonomous unit.

Not all segments of Czech public opinion favored a Pan-Slav solution of the Czech question, however. On the political right, conservative Catholic opinion had no wish to replace the Roman Catholic

Habsburgs with the Greek Orthodox Romanovs on Bohemia's throne. On the political left, the Czech Social Democrats regarded Tsarist Russia as the most powerful bastion of reaction and foe of socialism in Europe. They preferred, therefore, Austria-Hungary, despite all its faults. Finally, the Czech democratic left likewise had serious misgivings about the blessings of life under the tsars.

The principal spokesman of the liberal left was Thomas G. Masaryk (1850-1937), a professor of philosophy at the Czech University of Prague, who had won international renown as a scholar and great prestige and influence in the Habsburg empire, not so much as the leader of the small Czech Realist party which he represented in the Austrian *Reichsrat* (parliament) in Vienna but as the intellectual mentor and spokesman of all the Austrian Slavs. Almost alone among the Czech politicians, he took an active interest in the Slovaks of Hungary. He also took up the cudgels for the accused Serb and Croat politicians in the notorious Zagreb (Agram) Treason Trial of 1909.

Like most Czechs, Masaryk was originally under the spell of the legend of Russia — leader and protector of the Slavs. He visited Russia several times and carefully studied its life and culture. His book *Russia and Europe* (originally published in German, in 1911, and in 1919 in an English translation under the title *The Spirit of Russia*) became a standard study of pre-Marxian Russian thought. Although he had found much to admire in Russian culture, Masaryk lost faith either in the ability or the desire of the Russian tsars to free and lead the Slavs. He therefore did not advocate the break-up of the Habsburg empire before World War I, but only it reorganization along federal and democratic lines. Partly out of his profound belief in democracy and partly out of his concern for the fate of the Slovaks of Hungary, he did not base the Czech claims on Bohemia's historic state right but rather on the natural right of all peoples to choose their government.

The Slovaks of Hungary

Like the Czechs, the Slovaks experienced a national awakening toward the end of the eighteenth century but did not make comparable progress during the nineteenth century. Unlike Bohemia, Hungary did not undergo an industrial revolution until the very end of the nineteenth century. Consequently, the Slovaks did not experience the same upward social mobility engendered by industrialism as the Czechs had. They remained largely an agrarian people, with a large rural proletariat of landless peasants. Their industrial working class was small and their middle class was limited to a diminutive "intelligentsia," consisting of village priests or ministers, parochial school teachers, country doctors, and small-town lawyers. The Slovak national ethos was characteristically rural.

Nevertheless, Slovak cultural life was lively at the beginning of the nineteenth century. They produced a number of fine poets and scholars, e.g. Ján Kollár (1793–1852) and Pavel J. Šafárik (1795–1861), who contributed not only to the Slovak national awakening but also to that of the Czechs and the Austrian Slavs generally. Unfortunately, this promising growth was stifled by revolution and reaction that swept over the Habsburg empire from 1848 to 1859. Slovak political and cultural life briefly revived during the empire's period of constitutional experimentation in the 1860s. In 1861 the Slovak leaders met in Turčiansky Svätý Martin and drafted a memorandum for Emperor Francis Joseph, demanding Slovak administrative and cultural autonomy. Although the demands of the "Martin Memorandum" were not granted, the Slovaks were permitted to establish the Slovak *Matica* to serve as the clearing house of their culture and three *gymnasia* (classical high schools) to train their national intelligentsia.

This revival proved short-lived, however. By the Austro-Hungarian Compromise of 1867, Francis Joseph abandoned Hungary's national minorities for the sake of coming to terms with the Hungarian ruling class. Given a free hand, the Hungarian government adopted a policy of assimilating the minorities, the aim of which was to transform the polyethnic Hungarian kingdom into a national Magyar state. At the same time, it pursued singularly unenlightened social policies, the result — if not the intent — of which was to keep the country's masses, including the Magyar masses, poor, ignorant and obedient. One symptom of these policies was social emigration, which reached the proportions of a mass flight in the first decade of the twentieth century. Slovak emigration was directed principally towards the United States, in which eventually about one third of the Slovak people settled.

By the outbreak of World War I, the Slovak people were, as a conscious group, nearly on the point of extinction. Their cultural and political life had come to a virtual standstill. In these bleak circumstances, the Slovak leaders, typified by Svetozár Hurban Vajanský (1847–1916), poet and editor of the organ of the Slovak National party, pinned all their hopes on the great "White Tsar" of Russia to liberate them. Unlike the Czech Russophiles, however, the Slovak Russophiles hoped for outright Slovak incorporation into the Russian empire.

World War I

When World War I broke out, the Austrian government declared a state of emergency and for three years ruled dictatorially under the emergency paragraph 14 of the Austrian constitution of 1867. Even before the declaration of war, the government indefinitely prorogued the Austrian parliament and the diets of the Bohemian crownlands. As a wave of war hysteria swept over the empire, the government suspended trials by jury and extended the jurisdiction of military courts martial over civilians accused of committing antiwar or treasonous acts. The Hungarian government at Budapest did not dissolve the Hungarian parliament, in which the minorities had in any event only token representation, but cracked down on opposition groups just as hard as the government in Vienna did. Many Czech and Slovak leaders were arrested, while others were silenced by the simple expedient of being drafted into the army and sent to the front, even when they did not meet the physical requirements for military service. In this atmosphere of terror, the Czech and Slovak leaders fell silent and passively watched the war unfold.

As the notorious "Russian steamroller" lunged into Austrian Galicia and approached their lands, Czech and Slovak hopes for Russian liberation ran high. Kramář quietly passed the word to his followers to sit tight and wait for the Russians, who would, he thought, "do it for us alone." However, the battles of Tannenberg and the Marne in August and September of 1914 shattered the general assumption that the war would be short and decisive and dampened Czech and Slovak hopes of an early Russian liberation. Instead, the battles raised the prospect of a long-drawn-out war of attrition, to which the modern nation-staes of Western Europe (Britain, France and Germany) would quickly adjust but the antiquated multinational empires of Central and Eastern Europe (Austria and Russia) might not.

The possibility that Austria and Russia might break up under the strains of a total war convinced Masaryk that the Czechs and Slovaks should not wait for others to "do it for them," but must try to seize their destiny with their own hands. He believed that they themselves must take the initiative toward their liberation and employ all means at their disposal, political and military, to achieve it. They should not depend on Russian liberation alone but should also try to enlist the support of the western Allies. In September and October of 1914, he made two trips to neutral Holland to sound Western opinion out about support of the Czechoslovak cause. At that early stage of the war, the Allied powers had not yet formulated a common policy towards the Habsburg empire and its disaffected nationalities. Indeed, the western Allies were scarcely aware of the latter. Masaryk realized that a large-scale propaganda effort was necessary to engage their interest, but was encouraged by the steps already taken in this direction by Czechs and Slovaks living abroad.

Czechs and Slovaks Abroad

The first calls for Czechoslovak independence were actually sounded abroad. In Russia, Czech and Slovak colonists issued anti-Austrian declarations and petitioned the Russian government to permit them to form a voluntary Czechoslovak military unit to fight alongside the Russian army against Austria-Hungary for the liberation of their kinsmen at home. Historically, the Russian government had shown more interest in the Balkan Slavs than in the Austrian. After the outbreak of World War I, however, it took an interest in Czechoslovak aspirations. On August 18 the Russian high command issued the requisite orders for the organization of the volunteer Czech *druzhina* in Kiev, which proved to be the nucleus of the later famous Czechoslovak Legion in Russia. Two days later, Tsar Nicholas II received a Czech delegation in the Kremlin in Moscow. The delegation expressed to him the Czech desire that "the free and independent crown of Saint Wenceslas shine in the rays of the crown of the Romanovs."

The small colony of Czechs and Slovaks in France likewise demonstrated against Austria-Hungary after the declaration of war, and petitioned the French government to permit Czech and Slovak residents in France who were not French citizens to enlist in the French army. France had no traditional interest in the Czechs or Slovaks. It needed manpower, however. Accordingly, as early as August, 1914, the French high command created special units in the French Foreign Legion, which were reserved for alien volunteers who wished to serve against the Central Powers. One of these units was "Company *Nazdar,*" which was composed of Czech and Slovak volunteers. The first Czechoslovak volunteers in France were killed almost to the last man in the heavy fighting on the Western front in 1915–1916, but Company *Nazdar* was replenished with new volunteers, principally from the United States, and became the nucleus of the Czechoslovak Legion in France.

The United States was the home of the largest Czech and Slovak colonies abroad, numbering about 823,000 people according to the U.S. census of 1910. Before the war, while relations between individual Czechs and Slovaks in America were often cordial owing to their linguistic and cultural proximity (the Czechs and Slovaks can converse fluently, each using their own language, and understand each other), relations between the two communities as a whole were not especially close. Each community developed its own fraternal and cultural organizations, designed to promote their interests and welfare in the United States. Neither community organized any political or revolutionary movement designed to disrupt Austria-Hungary and free its compatriots in the "old country" from German or Hungarian domination, as the case might be. It was only the outbreak of World War I which rallied the Czechs and Slovaks in America and prompted them to formulate political programs for the liberation of their compatriots in Europe — even while the United States remained neutral until 1917.

The Bohemian National Alliance

At first, the Czechs and Slovaks in America pursued separate but parallel courses of action. On the Czech side, the first call for action was sounded by Jan Janák on August 12, 1914, in the Czech paper *Osvěta,* which he edited and published in Omaha, Nebraska. Assuming that the war would be short (as most people did at its outbreak), Janák looked to the peace conference and expressed the hope that it would "grant the Czechs independence or at least self-rule." He was aware that the great powers were not likely to do this without being prompted. Therefore he urged: "It is up to us living outside Austria to take the first step and send the Russian, English and French governments petitions or our personal representatives to press for Czech independence." His appeal concluded with a stirring call: "Long live the United States of Bohemia, Moravia, Silesia and Slovakia!" Thus, it was an American Czech who first publicly suggested not merely restoring the medieval Bohemian kingdom but establishing a modern Czechoslovak state.

In September, 1914, partly in response to Janák's appeal, the American Czechs founded the Bohemian National Alliance of America in Chicago to promote the cause of Czechoslovak independence. At first, the Bohemian National Alliance represented primarily the liberal, "free-thinking" element among the American Czechs, which looked to Prof. Masaryk for inspiration. In 1915, however, the Czech

American socialists and finally, in 1917, the Czech American Catholics also rallied to it.

The Slovak League

The American Slovaks were less divided than the Czechs but more cautious in their aims. The principal division in Slovak ranks was not political but religious; it was the division between the Catholic majority and Lutheran minority. Since 1907 the American Slovaks had in the Slovak League of America an organization which bridged the gap between the Catholics and Lutherans and represented the American Slovak community as a whole. In the spring of 1914 Count Michael Károlyi, the well-known Hungarian political leader, toured the United States. American press reports of his speeches revealed woeful ignorance on the part of the American press and public of conditions in Old Hungary. This started a discussion in the American Slovak press on the need for enlightening American and world opinion about the true nature and aspirations of the Slovaks. The outbreak of the war in Europe in July added urgency to the discussion. The outcome of it was the adoption of the "Memorandum of the Slovak League in America" in September, 1914. Like its historical predecessor in 1861, the League's Memorandum called only for Slovak administrative and cultural autonomy in Hungary. It was translated into five languages, printed in a handsome brochure and sent to the State Department and the belligerent embassies in Washington.

The Slovak League's program fell short of that of the Bohemian National Alliance. In December, 1914, the American Czechs took the initiative toward bringing the Slovaks around to their point of view. Their dialogue was eventually fruitful. At a joint meeting in Cleveland in October, 1915, the Bohemian National Alliance and the Slovak League concluded the so-called "Cleveland Agreement." It called for the "independence" and "union of the Czech and Slovak peoples in a federal union of states with full national autonomy for Slovakia," and provided the basis on which the two organizations thereafter cooperated.

The Czech Mafia

Although the initiative of the various Czech and Slovak colonies abroad was important, it alone could not assure Czechoslovak independence. This would depend primarily on the attitude of the Czech and Slovak peoples at home and the actions of their leaders. After his exploratory trips to Holland, Masaryk confidentially informed other Czech party leaders of his plans to launch an independence movement and proposed to them the formation of a secret committee to direct the movement. Before the committee was formed, however, Masaryk was warned, while on a trip to Switzerland in December, 1914, that the Austrian authorities had become suspicious of his activities and that he might be arrested on returning to Prague. He decided to stay abroad, therefore, and work for Czechoslovak independence openly. The task of organizing and initially directing the secret committee, which after the war became known as the "Czech Mafia," fell to his disciple Eduard Beneš (1884–1948), a young sociology professor in Prague.

From Zürich Masaryk went to Rome, which was at that time still a neutral capital, and from Rome to Allied Paris and London. In Paris, he encountered his former student in Prague — Milan R. Štefánik (1880–1919), a young Slovak astronomer, scientist, meteorologist, traveler, and, since the war, officer in the French army air force. Through his acquaintances in Paris, Štefánik arranged an interview for Masaryk with the French premier, Aristide Briand. In London, Masaryk was introduced to Prime Minister Herbert Asquith by Henry Wickham Steed, the political editor of the *Times* and former correspondent in Vienna, who was well acquainted with the problems of the Habsburg empire. The French and British premiers listened courteously to Masaryk's pleas for Czechoslovak liberation, but remained noncommittal. Being primarily concerned with the defeat of Germany, they did not wish to prejudice the possibility of detaching Austria-Hungary from the Central Alliance by committing themselves to the cause of its disaffected nationalities.

In May, 1915, Kramář and several other members of the Czech Mafia were arrested. Kramář and his close collaborator Alois Rašín (1867–1923) were tried and convicted of treason and sentenced to death. Their execution was postponed, however. In September, fearing arrest, Beneš fled into exile. He joined Masaryk in Paris, thereafter becoming his lifelong, closest collaborator.

The Czechoslovak National Council

On November 14, 1915, the Paris exiles formally launched the Czechoslovak movement for independence by issuing, in the name of the "Czech Foreign Committee," a declaration demanding the establishment of an independent Czechoslovak state. In 1916, the Czech

Foreign Committee was transformed into the Czechoslovak National Council of Paris. Masaryk became its chairman; Štefánik and Josef Dürich, vice chairmen; and Beneš, secretary-general. Dürich, a Czech Agrarian party leader and recent exile, went from Paris to Russia, where he allowed himself to be used by pro-Russian, conservative, antidemocratic, and anti-Western exiles to challenge Masaryk's leadership and set up a rivaling national council in Petrograd. After the outbreak of the Russian Revolution in March, 1917, the Petrograd National Council was dissolved, however, and the unity of the Czechoslovak independence movement under Masaryk's leadership was restored.

The three leaders of the Paris National Council — Masaryk, Štefánik and Beneš — proved to be, despite the differences in their age, background and character, an exceptionally effective team. Prof. Robert W. Seton-Watson, a Scottish historian and close student of Habsburg affairs, remarked about the Czechoslovak liberation triumvirate that they were always in the right place at the right time. Indeed, they were. Until 1917, Masaryk felt that because of his knowledge of English and the Anglo-American world he could best serve the Czechoslovak cause in London. Thanks to Seton-Watson's recommendation, he was invited to lecture at King's College, which provided him with a suitable platform from which to address himself to the British public. After the fall of the Russian monarchy in 1917, which opened the gates of Russia previously closed to him, he hastened to Petrograd and secured from the Russian high command an agreement to expand the Czechoslovak Legion and elevate it in status to an autonomous Allied cobelligerent army. When the November (Bolshevik) Revolution and the Soviet withdrawal from the war ended its usefulness on the Russian front, Masaryk accepted the French proposal to place it under French command and transfer it to the Western front. Then he set out via Siberia and Japan for the United States, where he hoped to influence President Woodrow Wilson, who had emerged as the principal Allied spokesman.

As a French officer, Štefánik was able to travel easily between Allied countries and found many doors which were closed to Masaryk and Beneš open to himself. He became, therefore, the National Council's principal troubleshooter. In 1916, he hastened to Russia to repair the damages caused the Czechoslovak movement by Dürich's inopportune action. After the American declaration of war on Germany in 1917, he went to Washington and secured official American permission to recruit volunteers for the Czechoslovak Legion in France from among Czech and Slovak immigrants. In 1918, he went to Rome and secured Italian recognition of the Czechoslovak movement and permission to organize the Czechoslovak Legion in Italy. Later in the year, he travelled through the United States and Japan to Siberia to take charge of the Czechoslovak Army, which had become entangled in the Russian Civil War and Allied Intervention and had failed to reach the Western front. As Secretary General of the Czechoslovak National Council, Beneš remained in Paris and directed its affairs. In 1917, he scored a brilliant diplomatic success by persuading the French government to include Czechoslovak liberation among the Allied war aims, which for the first time raised the Czechoslovak question from the arena of internal Habsburg politics to the international level.

Allied War Aims

In a note to President Wilson on January 10, 1917, the Allies included among their war aims "the liberation of Italians, of Slavs, of Rumanians, and of Czechoslovaks from foreign domination," which implied that they intended to dismember the Habsburg empire. The young Emperor Charles, who had succeeded Francis Joseph in November 1916, was thoroughly frightened. To counter the implied threat of dismemberment, he moved to conciliate the disaffected nationalities by rescinding martial law for civilians in noncombat zones, convoking the *Reichsrat* in May, and amnestying condemned national leaders, including Kramář and Rašín, in July.

Abroad, Emperor Charles put out peace feelers to the Allies, which revived their hopes to detach Austria-Hungary from Germany and isolate the latter for the kill. In secret negotiations, stretching from March, 1917, to April, 1918, the French, the British, and the Americans, each in turn, tried to persuade Vienna to conclude a separate peace on the basis of preserving the Habsburg empire essentially intact. The negotiations ultimately failed, because Vienna would not or could not break with its powerful German ally. The Allies then reluctantly turned to the alternate policy of encouraging the disaffected nationalities of the Habsburg empire to revolt by holding the prospect of independence out to them.

In April, 1917, when President Wilson recommended that Congress declare war on Germany, he counselled it against declaring war on Austria-Hungary because it would disrupt his secret negotiations with Vienna. In December, 1917, when he recommended a declaration of war on Austria-Hungary, he hastened to give Vienna an assurance against dismemberment. A month later, in his famous Fourteen Points address, he called only for "the freest opportunity of autonomous development" for "the peoples of Austria-Hungary" (Point Ten), not for their independence. It was not until April, 1918, that he definitely abandoned his hopes of weaning Vienna away from Berlin and endorsed the Czechoslovak and Yugoslav movements for independence.

Masaryk, who arrived in the United States and toured the American cities at that time, was given a cordial reception in official American circles and a hero's welcome by the American Czechs and Slovaks. In the famous "Pittsburgh Agreement," concluded in his presence on May 30, the American Czech and Slovak organizations reaffirmed their common political program. After the war, the Pittsburgh Agreement became an object of bitter controversy, but at the time of its conclusion it expressed the good will and confidence that the two peoples then felt towards each other.

Declaration of Czechoslovak Independence

In July, 1918, the fortunes of war definitely inclined in favor of the Allies on the Western front. The prospect of an Allied victory encouraged the Czech Mafia to come out into the open and form the Czech National Committee in Prague. Kramář became its chairman and Antonín Švehla (1873–1933), an Agrarian party leader and future Czechoslovak prime minister, its moving spirit. On October 4, frightened by the Allied advances on the fronts, Vienna offered to surrender on the basis of the Fourteen Points. To implement the demand of Point Ten for the autonomy of his peoples, Emperor Charles issued, on October 16, a manifesto announcing the federalization of the Habsburg empire. Because of violent opposition of the Magyars to the manifesto, Hungary was expressly exempted from its provisions. This weakened it. Even if Hungary had been included in it, however, it is doubtful whether at that late date the manifesto could have saved the empire. In any event, on October 16 President Wilson declined Vienna's offer of surrender on the basis of the Fourteen Points, on the ground that since the issuance of that program he had committed himself to support Czechoslovak and Yugoslav independence. On October 27 Vienna bowed to the inevitable and unconditionally asked for an armistice. This became the signal for the dissolution of the Habsburg empire.

On the following day, October 28, the Prague National Committee issued its first "law": "The independent Czechoslovak state has come into being." On October 30, unaware of the Prague declaration, Slovak political leaders met at Turčiansky Svätý Martin, formed a Slovak national council, and declared for Slovak independence from Hungary and union with the Czechs. The multiple tasks of setting the new state afloat took months to complete. On November 13 the Prague National Committee adopted a provisional constitution, declaring Czechoslovakia to be a parliamentary republic. Under this constitution, a provisional ("revolutionary") national assembly met in Prague on the following day, elected Masaryk (then still in Washington) president and invested a cabinet with Kramář as prime minister, Beneš as foreign minister, and Štefánik as war minister. By the time Masaryk returned to a hero's welcome in Prague, on December 21, Czechoslovakia was a viable state.

SUGGESTED READING

Beneš, Eduard, *My War Memoirs* (New York, 1928).

Kerner, Robert J., ed., *Czechoslovakia* (Berkeley, 1940).

Mamatey, Victor S., and Radomír Luža, ed., *A History of the Czechoslovak Republic, 1918–1968* (Princeton, 1973).

Mamatey, Victor S., *The United States and East Central Europe, 1914–1918: A Study in Wilsonian Diplomacy and Propaganda* (Princeton, 1957).

Masaryk, Thomas G., *The Making of a State: Memories and Observations, 1914–1918* (London, 1927).

Opočenský, Jan, *The Collapse of the Austro-Hungarian Monarchy and the Rise of the Czechoslovak State* (Prague, 1928).

Pergler, Charles, *America in the Struggle for Czechoslovak Independence* (Philadelphia, 1926).

Perman, Dagmar, *The Shaping of the Czechoslovak State, 1914–1920* (Leiden, 1962).

Seton-Watson, Robert W., *A History of the Czechs and Slovaks* (London, 1943).

———., *Masaryk in England* (Cambridge, 1943).

Thomson, S. Harrison, *Czechoslovakia in European History* (Princeton, 1953).

Josef Anderle

THE FIRST REPUBLIC, 1918–1938

The Provisional Administration. The political system by which the country was administered between 1918 and 1938 came to be called the First Republic. It was defined by the Provisional Constitution, which had been worked out by the Revolutionary National Committee on November 13, 1918, as a parliamentary republic such as there was in France, in contrast to the presidential democracy that existed in the United States of America. That meant that the balance of power in the new state was tipped in favor of the parliament, the National Assembly, which was endowed not only with a near monopoly in the legislative branch of the government, but also with considerable control over its executive and judicial branches. The president of the republic was the head of state and its chief representative in foreign relations, held supreme command over the armed forces, appointed ambassadors, generals, ministers and most of the chief justices, but virtually all of these affairs required the prior approval of the government, which in turn was collectively responsible to the parliament, over whose activities the president held only a faint veto power that could be easily overturned. Tomáš G. Masaryk, who was elected the first president of the republic by the National Assembly on November 14, 1918, was more in favor of the more stable and efficient American system of government, but the bad experiences of the Czechs and Slovaks with the absolute monarchs of the past influenced their leaders to opt for a strong parliament as a guarantor of the general will of the people, the ultimate source of state power.

But the Provisional National Assembly of Czechoslovakia, inaugurated on November 14, 1918, did not properly reflect the general will of the population of the state. Since the frontiers of the republic had not been fully determined and secured, and since it was impossible to hold parliamentary elections in the first months after the war, the composition of its membership was chiefly determined by the last parliamentary elections held in the Czech lands in 1911. In Slovakia such elections, owing to an underdeveloped political life and grossly unjust election laws and practices in prewar Hungary, had produced only three Slovak representatives in the Hungarian parliament; therefore the Prague National Committee appointed forty deputies of the Slovak people to the National Assembly arbitrarily, mostly on the advice of Dr. Vavro Šrobár, the sole member of the Committee from Slovakia. Fourteen additional seats were subsequently reserved for the Slovaks in the National Assembly so that they were represented by fifty-four deputies, while two hundred and fourteen deputies represented the more populous Czechs. Originally it was planned to include leaders of the German minority and other nationalities of the state in the National Assembly, but these plans were postponed because they had refused cooperation or because the fate of their districts had not been determined for a long time. This gave some members of these nationalities an excuse to claim later that they were not bound by any allegiance to the state since they had not participated in its formation. But the National Assembly did represent all Czechs and Slovaks, who comprised more than two thirds of the population and worked fairly harmoniously on the solutions of common problems for the benefit of all ethnic groups.

The government, too, represented only the Czechs and Slovaks at this time. Aside from Prime Minister Karel Kramář and Foreign Minister Eduard Beneš, it included such other experienced Czech politicians as Antonín Švehla, Alois Rašín, Gustav Habrman, František Soukup, Václav Klofáč, as well as two Slovaks, Vavro Šrobár and Milan Štefánik. The latter, however, did not have much chance to contribute to its efforts, for he died early in May, 1919, in an airplane crash as he was returning home from his wartime activities abroad. In July, 1919, after communal elections in the Czech lands had revealed a marked shift of public opinion toward the left, the composition of the government was adjusted accordingly and Dr. Kramář, who had led the conservative National Democratic Party, yielded the premiership to a Social Democratic Party leader, Vlastimil Tusar.

The parliament and the government faced a number of challenging tasks. They had to secure the desired boundaries of the state, develop a central administration and establish its authority in the whole territory of the state, meet the most essential needs of the population, which had suffered grievously from the consequences of the long and exhaustive war, and give the country a permanent constitution.

The State Boundaries. The desired boundaries of Czechoslovakia were essentially secured through peace treaties with Germany (June, 1919), Austria (September, 1919), and Hungary (June, 1920). The boundaries of the western part of the country generally coincided with the historic boundaries of the Czech lands and were recognized by the Allies in November, 1918. They included districts inhabited by large numbers of Germans, who wished to join Austria, or — along with Austria — Germany, but none of these wishes were granted to them by the Peace Conference, mostly because they were either impractical, or incompatible with Allied interests. The German districts of the Czech lands did not form large enough areas, but were strung along the borders of these lands in a narrow ring, separated from each other, as well as from Austria and Germany, by Czech-inhabited areas and high mountain ranges; and they would have increased intolerably the economic and strategic potentials of defeated Germany if they had been attached to it. The Czechoslovak government also claimed these territories on historic and defensive grounds. They had always been integral parts of the lands of the Bohemian Crown, and the old boundary line, as it ran on the crest of the mountains encircling these lands, provided the state with a frontier that could easily be defended. The preservation of the natural economic unit, which the German border regions formed with the Czech interior, was another argument that was appreciated even by many Germans. Since their reluctance to remain within the historic borders with the Czechs was principally influenced by their fear of Czech revenge for the bad treatment the Czechs had often experienced in the German-dominated Habsburg Empire, the Czechoslovak government undertook a special commitment in September, 1919, in a separate "Minority Treaty" with the Allies, by which it guaranteed Germans and other minorities full exercise of their civil rights and full development of their cultural heritage. Encouraged by the Allied attitude, Czechoslovak authorities occupied the German districts by the end of December, 1919, without any significant resistance.

Establishing the boundaries of Slovakia proved to be more difficult. Slovakia had never been an independent state or administrative unit, and so a claim to historic boundaries could not get too far. Most Slovaks lived in the mountainous area north of the Danube and Ipeľ rivers, which could not be made into a viable province, unless it were joined with the fertile plains along these rivers that would also provide it with a defensible frontier. But these areas were inhabited mostly by Hungarians who wished to remain in Hungary. Slovak historians claimed that these districts had been originally settled by the Slovaks as well — Hungarians being relative newcomers there — but these claims were always sharply refuted by Hungarian writers on the subject. In fact, the Hungarian government wished to retain not only these disputed plains, but the Slovak-inhabited territory as well. It tried to accomplish this by offering the Slovaks autonomous government, while an adventurer in its pay attempted to organize eastern Slovakia as an independent republic and the Hungarian army frustrated efforts of the Prague government to establish its control in the area. But these endeavors failed in January, 1919, when the Allied powers accepted the two rivers, plus a line extending further eastward to the upper Tisza river, as a provisional boundary and demarcation line. However, in May of that year, after a Bolshevik government had proclaimed Hungary a Soviet republic, the Hungarian army invaded parts of Slovakia again, making it possible for a few Slovak Bolsheviks to proclaim Slovakia a Soviet republic too. But this effort also collapsed within a month, owing to diplomatic interventions of the Allies, military countermoves by Czechoslovakia and Romania, and an anti-Hungarian movement in Hungary itself. The Peace Conference then acknowledged the provisional boundary as the final one and a new Hungarian government accepted it too, although the relevant peace treaty was not signed until June, 1920.

The Peace Conference also added to Czechoslovakia the small territory of Ruthenia adjacent to eastern Slovakia. It was inhabited by a Slavonic people related to the Ukrainians, whose representatives in America had concluded a tentative agreement regarding such a union with Tomáš Masaryk in October, 1918. The Hungarians remained in control of Ruthenia, however, and in May, 1919, Hungarian Bolsheviks tried to set it up as a Soviet republic too. But a Central Ruthenian Council approved the union with Czechoslovakia with the stipulation that Ruthenia would obtain autonomous self-government, and this arrangement was accepted by the Allies as well as Hungary. The territory again included not only the mountainous area in which the Ruthenians lived but also the plains along the upper Tisza river, which were inhabited mostly by Hungarians. The rights of these Hungarians, as well as those in Slovakia, were guaranteed by the Minority Treaty mentioned above.

The Peace Conference did not settle the boundary of Czechoslovakia with Poland. Although most of the old frontier between Poland and Hungary was accepted as the new Polish-Czechoslovak border by both countries without any quarrels, serious disputes had developed over the small territories of Těšín, Orava, Spiš and Javorina. Eventually these disputes were resolved by a series of compromises arranged by the ambassadors of the Allied powers in 1920 and 1923.

Population Structure. After the frontiers of Czechoslovakia had been definitely fixed, the Czechoslovak Republic emerged on the map of Europe as an elongated tract of territory stretching almost 600 miles from west to east between Germany, Austria, Hungary, Romania and Poland; covering an area of about 56,000 square miles inhabited by nearly 13.4 million people, as the census of 1921 revealed. It was a small state, but not among the smallest in Europe. Among the thirty-two states of that continent, Czechoslovakia ranked thirteenth in area and ninth in population. It was considerably larger than such states as Ireland, Belgium, the Netherlands, Switzerland, Hungary, Portugal, and most Baltic and Balkan countries. If it had been included in the United States of America at that time, it would have ranked twenty-sixth in area, comparing favorably with such states as Illinois or Wisconsin, but first in population, outnumbering New York and California.

Since the state boundaries of Czechoslovakia did not exactly follow the ethnic boundaries of the Czech and Slovak peoples, the population of Czechoslovakia included several national groups. The largest among them were the Czechs, who numbered a little over six million people in 1921. The closely related Slovaks numbered a little less than three million. Together they accounted for almost 65.5 per cent of the total population. Among national minorities the largest were the Germans, who counted 3.1 million, or about 23 per cent of the total; they lived mostly in the Czech lands, where their share in the population was almost 30 per cent. The next largest minority group were the Hungarians, whose number came close to 750,000, or nearly 5.6 per cent of the total; they lived in Slovakia and Ruthenia, where they made up 21 and 17 per cent of the population, respectively. The Ruthenians counted almost half a million people (3.5 per cent) and the Poles only 76,000 (0.6 per cent). Czechoslovak laws made it possible for citizens of Jewish faith to register as Jewish nationals and a little over 180,000 did so (1.3 per cent). The remainder of the ethnic structure of Czechoslovakia was composed by still smaller groups of a few other nationalities. (Citizens of foreign countries, who counted almost 250,000, are not included in this survey.) In general, this picture remained fairly stable in the next two decades. All ethnic groups except for the Hungarians, increased in absolute numbers, but only the Czechs and Slovaks registered important gains in relative numbers. By 1938, the Czechs and Slovaks counted over ten million people, or about 68 per cent in the total population. These changes were due partly to a decline in emigration among the Czechs and Slovaks, a greater increase in their birthrate and national consciousness, and even some reemigration. Smaller natural increases, larger emigration (mostly across the Czechoslovak borders to Germany, Austria, Hungary and Poland), as well as some assimilation with Czechs and Slovaks, accounted for the smaller increases in the numbers of the minorities and some decline in their share of the total population of the state.

The new state was endowed with considerable economic resources. It had large areas of arable land which produced almost all the food it needed, and large deposits of industrial raw materials which encouraged development of a prosperous industry and lively commerce, especially in the Czech lands. This was reflected in the occupational structure of the population, which showed a fair balance between people engaged in agriculture on one side, and industry and crafts on the other, each of them employing a little over one third of the population, while the remainder were divided among other categories of economic activity, most of which could be put under the heading of services. In subsequent years a tendency developed toward a decline of agricultural population in favor of people employed in industry, commerce and the civil service, as the country advanced along the path of industrialization and urbanization. But in terms of wealth distribution the social structure of Czechoslovakia was less balanced, although it did not exhibit the sharp inequities of some other countries of Europe. The core of the Czechoslovak society rested on the related strata of the lower middle class and upper low class, which jointly accounted for almost 55 per cent of the population in 1921. This group included mostly independent peasants and artisans, small tradesmen, skilled workers, and the lower ranks of civil service and other service groups. The lower layer of the low class, however, was still a formidable group, the largest single social stratum of the country, in fact, as it accounted for almost 39 per cent of the population. These people were small peasants or landless agricultural laborers, unskilled workers and the unemployed. However, the upper layer of the middle class accounted for only 5.7 per cent of the population, while the entire high class for less than one per cent. These people included large landowners, industrial and commercial entrepreneurs, bankers, free professions, and higher ranks of civil and military service. Nobility was almost absent from this picture. Native Czech nobility was largely destroyed, expelled or denationalized during and after the Thirty Years War of the seventeenth century, and although the foreign nobility, mostly German, which took its place, was often very wealthy, it was inconsequential in numbers. Slovak native nobility had been assimilated within the Hungarian nobility long before the establishment of the Czechoslovak Republic.

The religious composition of the Czechoslovak population was dominated by the Roman Catholics, who accounted for 76.3 per cent of the total population in 1921. However, in subsequent years they registered a loss of almost two per cent due to a large scale desertion of the Catholic Church by many Czechs who were scandalized by the past links of the higher clergy, often of German origin, with the Habsburg dynasty. Some of these people formed a new denomination, the so-called Czechoslovak Church, which first tried to find an independent position between the Catholics and Protestants, but later inclined toward the latter; by 1930 it counted almost 800,000 members (5.4 per cent of the population). The Protestants were numerically small (7 per cent), but often excelled over the Catholics and occupied influential positions in the society and state. Christians of the eastern rites were organized in the Greek Catholic Church (3.9 per cent) and the smaller Orthodox Church (0.5 per cent); they lived mostly in Ruthenia and eastern Slovakia, where they first attracted mostly the people who identified with the Ukrainian or Rusyn nationality, while people who considered themselves Russians adhered to the latter. The Jewish religion was practiced by about 350,000 people (2.6 per cent).

As already indicated, individual provinces often differed considerably in their population structure in most of these aspects. Almost half of the poulation of Czechoslovakia lived in the westernmost province, Bohemia, which was also the largest in area and most advanced economically and culturally. The other two Czech lands, Moravia and Silesia, though considerably smaller, did not lag much behind Bohemia in their economic and cultural development, but Slovakia and Ruthenia, long neglected by semi-feudal Hungarian governments, subsisted almost exclusively on an agrarian, or even pastoral economy, and illiteracy among the adult populations there was as high as 35 and 57 per cent, respectively, before the war. It was only through the energetic efforts of the Czechoslovak government that illiteracy was reduced to 8 and 31 per cent in Slovakia and Ruthenia respectively by 1930. In the Czech lands the smaller prewar illiteracy rate of 2.5 per cent was reduced to 1.3 at the same time.

Emergency Social Reforms. Surprisingly enough, the new state's most urgent problems did not arise from the national composition of its population, but from the social and economic divisions inherited from the past, especially from the difficulties generated by the world war. They were aggravated by revolutionary propaganda from Bolshevik Russia as well as from Hungary and several areas of Germany and Austria, where Bolshevik or other radical regimes were established too, however briefly, preaching a world revolution.

A long-resented burden was the unequal distribution of land. According to the Austrian census of 1896, about 600,000 peasant families in the Czech lands owned less than 300,000 hectares, approximately as much as was held there by only three noble families. Indeed, the 151 large landlords of the Czech lands, mostly German nobility, possessed nearly 1.5 million hectares. Most peasants could earn a livelihood for their families only by working on the large estates for long hours and low wages. In Slovakia and Ruthenia, where Hungarian nobility almost monopolized land ownership, the conditions were even worse. A similar situation existed in industry and commerce, two thirds of which were owned by German or Hungarian entrepreneurs and bankers. Workers had to work for as long as thirteen hours a day for wages that fell sixty per cent below the prewar levels, owing to a war-fanned inflation, which increased the prewar prices of food and other commodities often as much as twenty times. The workers' position was further aggravated because many of the consumer industries were converted to serve war production and re-conversion to peacetime production often meant a long pause in their operation. Such pauses were also caused by the disruption of international trade, on which Czechoslovak industry depended for supplies of capital or raw materials and for markets for finished products. As a consequence, many workers lost their jobs and returning soldiers inflated the numbers of the unemployed to a figure of more than 350,000 (about 8 per cent of the labor force) by 1919.

The desperate and near starving peasants and workers demanded immediate reforms and often took matters into their own hands by taking over vacant estates or idle factories. Bolshevik-inspired agitators called for a complete social revolution and the establishment of a Soviet regime in the country. Because Czechoslovak labor was a large and well organized force, prospects for such a revolution were taken quite seriously and the government on several occasions considered suggestions to call in the French army to help maintain order. But revolution did not break out and the government was able to contain the revolutionary fervor with its own forces because it was wise enough to respond to the needs of the people by radical social and economic reforms. Many lawmakers viewed these reforms as a revolution of their own, and the Provisional National Assembly came to be called the Revolutionary National Assembly as well.

In November, 1918, the National Assembly indicated its wish to carry out a radical land reform and sequestered all large estates for that purpose. After feverish work the land reform was enacted in April, 1919. It limited large estates to 150 hectares of arable land each (plus 100 hectares of forests and other non-arable land) and marked the released land (almost 32 per cent of all land) for distribution among the peasants, who had none or too little of it, although some of the land, especially forests, was nationalized and operated by the state. The original owners of the sequestered land were compensated by the beneficiaries of the reform, who received easy credit from the government for that purpose. Although the land reform took more than ten years to materialize fully and was flawed by favoritism and other deficiencies, it was the most comprehensive and equitable land reform in postwar Europe and was proudly referred to as "the greatest act of the republic" by President Masaryk. It certainly stabilized the fluid situation in the countryside.

Reforms in the field of industry and commerce were not achieved by such a single comprehensive act, but a whole series of specific measures that occupied the government and the National Assembly for several years. But in December, 1918, the working day was cut to eight hours and from then on a number of other acts established or improved workers' insurance against sickness and accidents, provided for their professional training and education as well as their vacations and pensions, and offered them protection against unjust dismissal and support during unemployment. Execution of some of these policies was left in the hands of workers' unions and workers' committees received the right to participate in the decisions of the plant managements affecting the health and welfare of their fellow workers. The National Assembly even met some of the workers' demands for nationalization of essential services, such as railroads, telephones, telegraphs and postal services,if they had not been under public control before, while municipal authorities were permitted to take over the operation of local utilities, such as water, electricity and gas plants, or local transportation. Most other industrial and commercial enterprises were left in private hands. But foreign owners and investors were compelled to relinquish the control of these enterprises to Czechoslovak citizens, companies and institutions, which benefited the Czech and Slovak middle class. Drastic laws against the black market eliminated profiteering, and a sharp devaluation of the Czechoslovak crown, along with other monetary and fiscal reforms, terminated the wartime inflation. The harsh burdens that the reforms initially imposed on the population angered many people of small income and incited a Communist youngster to assassinate Finance Minister Rašín in January, 1923, but the reforms spared Czechoslovakia the disastrous ravages that neighboring countries suffered from run-away inflation for years, and facilitated Czechoslovakia's entry into international trade, on which its economy depended heavily. Like the land reforms, these measures went further than similar policies of other countries of Central and Western Europe, and although they did not satisfy the workers completely, they went a long way toward winning their allegiance to the new state.

The Constitution of 1920. The Provisional and Revolutionary National Assembly also became a Constituent National Assembly, when it fulfilled its last task and gave the country its permanent constitution on February 29, 1920. In general, the new constitution confirmed the provisional one in its basic features, but expanded its brief language and filled in various gaps.

Ideologically, it defined Czechoslovakia as a democratic state, modelled on western countries, especially France, but often went beyond these models to respond to the contemporary requirements of democratic societies, certain particular needs of the Czechoslovak people and state, and various commitments undertaken in the peace treaties. Thus it guaranteed Czechoslovak citizens the customary rights and freedoms of western democracies, but added some new ones, e.g. the right to work and social insurance, as well as women's suffrage, and protection of marriage, motherhood and family. It promised an autonomous administration to Ruthenia (called Sub-Carpathian Russia), granted the national minorities equality with the Czechs and Slovaks, and guaranteed free development of their cultural institutions, as well as public support for them.

Administratively, the new constitution designated Czechoslovakia more clearly as a parliamentary democracy, guided primarily by a popularly elected National Assembly. The National Assembly was divided this time into two houses, of which the larger Chamber of Deputies (300 seats) prevailed over the Senate (100 seats). The National Assembly continued to enjoy dominance in legislative initiatives and control over the executive and judiciary branches of the government. It elected the president and confirmed the government appointed by him, but could overthrow both, the first by impeachment and the second by a vote of no confidence. The executive power was shared by the president and the government, but the latter prevailed over the former. The president was the head of the state, but not of the administration. Although his position toward the parliament and government was strengthened a little, he still remained responsible to the government, and the government was responsible to the parliament. He was given only a small part in the direct administration of the country, which was the realm of the government. The government was a collective team, which deliberated and acted jointly and bore collective responsibility for the executive branch. Only in emergencies could the government escape the control of the parliament, suspend civil rights and rule by executive orders, but then only for brief periods, after which all of its actions had to be submitted to the parliament for scrutiny and approval.

The judiciary system, which Czechoslovakia had inherited from Austria, was essentially good, fair and modern and the Provisional Constitution left it unchanged. But the Constitution of 1920 improved it further, as it rigidly separated ordinary courts from administrative agencies, decentralized the system to make courts more accessible to the people, safeguarded the independence of judges by additional provisions, and extended the jury system to new domains. Secret court proceedings were totally abolished, courts-martial were provided only for the case of war, and other special courts could be established only by law for specified cases and periods. The highest judicial authorities established by the constitution included the Supreme Court, which served as a court of appeals in civil and criminal cases; the Supreme Administrative Court, which adjudged violations of civil rights and conflicts between individual branches of the government; the Constitutional Court, which examined the constitutionality of legislative acts; and the Electoral Court, which handled complaints concerning irregularities in electoral processes and verified the results of elections.

Local government, also untouched by the Provisional Constitution, proved to be a difficult problem for the authors of the Constitution of 1920. As we have seen, Czechoslovakia was composed of territories, which in the past had been ruled by different codes of law, and in which ethnic and political interests sometimes pulled in opposite directions. Faced with such a situation, the authors of the constitution saw the state's task as first to unite these territories before granting them meaningful self-government. As a result of these concerns the new constitution, along with a special amendment passed on the same day, gave Czechoslovakia the form of a highly centralized state, again resembling the French Republic. The Czech lands with their provincial

parliaments and governments (diets and committees) were abolished
and divided, as were Slovakia and Ruthenia, into smaller territories
(counties), which were supposed to be administered by their own
governments, but under the close guidance and control of the central
government. However, this arrangement ran so much against local
traditions and preferences that it was never fully put in force, and in
1927 it was completely replaced by a new system which restored the
provincial parliaments and governments in the Czech lands, and introduced them in Slovakia and Ruthenia. But the central government did
not relinquish its control over them. Their jurisdiction was limited to
adjusting laws and regulations of the central authorities to local needs
and conditions, and government agents and agencies (provincial presidents and provincial executives) checked them still further at their
localities. In fact, the central government even appointed one third of
the membership of the provincial assemblies, so that they were only
partially elected by the local population. A similar system was introduced in the smaller units of the local government, the districts. Only
on the lowest levels, in local communities, was the government completely in the hands of, and elected by the local population. Although
this system was at first generally received with greater favor by the
people, pressures for another reform of local government built up
again in the 1930s, especially in Slovakia and the minority districts, so
that in 1938 the central government expressed its willingness to withdraw its agents from the local governments and expand their scope. In
that year also Ruthenia received the promised autonomy, although its
local government had from the beginning been supervised by a local
governor, appointed by the president of the republic. Silesia was
merged with Moravia in 1927, because its small size did not warrant
maintenance of a separate administration.

    Political Parties. The most decisive factor in Czechoslovak politics,
however, was not the president, the government, or even the parliament as a whole, but the political parties. The Czechoslovak political
system faithfully reflected the ethnic, social and economic structure of
the Czechoslovak population and even its religious and regional differences. This was evidenced in the multiplicity of the political parties
and their programs. There were distinct parties that represented the
agrarian population and interests, workers, small business and big
business. There were parties that represented the Catholics and virtually all shades of political ideology — except for the defaced
monarchism — from the Communists on the extreme left to the Nazis
and Fascists on the extreme right. Many of these political trends split

along national lines. Thus there was originally not only a Czech Agrarian Party, but also Slovak, German, Hungarian and Ruthenian parties; not only a Czech Social Democratic Party, but also a Slovak, German and Hungarian party; and not only a Czech but also a German National Socialist Party; and not only a Czech but also a German Tradesmen Party. Fervent nationalism was advocated by the Czech National Democratic Party, the German and Hungarian National Parties and a United Polish Party. Catholics, who felt the need to defend the interests of their church by political means, were organized in separate Czech and Slovak People's parties, and separate German and Hungarian Christian Social parties. Even the Communists originally organized in separate national parties. But a similar name or class base did not always denote a similar ideology. Thus the German National Socialist Party quickly drifted away from its genuinely democratic namesake, the Czech National Socialist Party, and developed into a totalitarian Nazi Party, as did the German National Party, while the Slovak People's Party soon left its sister Catholic parties in the center of the political spectrum and moved close to the extreme right. Some of the parties merged in subsequent years: Czech and Slovak Social Democrats in 1919, Czech and Slovak Agrarians in 1922, and all Communists in the fall of 1921. The Czech National Democrats formed a Party of the National Union in 1935 with several smaller parties of the right.

Nevertheless, confronted with this picture, a reader used to the familiar Anglo-American two party system may have a vision of complete chaos. Indeed, as many as thirty parties often competed for the votes at the polls and — with many following opposing goals — it seemed that the country was ungovernable. In fact, in neighboring countries, where population structures were often simpler, similar divisions led to a political impasse, which encouraged one group or another to suppress its competitors and establish its own rule over them. In Czechoslovakia, however, the seemingly unmanageable chaos was managed and the democratic regime was preserved.

Yet this is not difficult to explain. As we have already seen, Czechoslovakia possessed considerable economic resources, which made for a less inequitable distribution of national wealth and, consequently, for a more balanced social and political system. For the same reason class consciousness was not developed too sharply among the Czechoslovak population, and such class conflicts that had exploded in the country sporadically, despite their alacrity, did not result in rifts that could not be bridged over by reasonable leaders on both sides. Also, since these conflicts often had nationalistic undertones (Czech and Slovak peasants and workers often confronting German and Hungarian landowners and industrialists), they were considerably dampened in passion and number by the nationalization ("nostrification") of foreign ownership of excessive properties in land and industry and attendant concessions for the peasants and workers of all nationalities.

The ethnic divisions among the Czechoslovak population seemed potentially even more destructive, but were bridged over successfully too. The majority of the Czechs and Slovaks, fairly satisfied in their national aims, knew very well that they had everything to lose if they could not get along among themselves and win some cooperation from the minorities. Although the separation of Slovakia from Hungary had unfavorable consequences for the economy of the province that were never overcome satisfactorily, the Slovaks advanced rapidly in other fields, especially in national consciousness, education, and political organization, and were given an increasing share in the political and economic life of the state. It was never enough, especially fast enough, for the Slovak People's Party of Father Andrej Hlinka, which was the largest party in Slovakia and saw the best and quickest guarantee for Slovak advancement in the establishment of Slovak autonomy; most other Slovak parties were satisfied with the gradual progress directed by the central government, in which they were always represented. Similar attitudes had developed among the Ruthenians, where economic and cultural conditions were even more difficult and where more work had to be accomplished before the promised autonomy could be fully realized there. Although the attitudes of the Czechs and Slovaks toward the German and Hungarian minorities, who had once totally dominated them, was never free from justified complaints, it was repeatedly attested by foreign observers that the Czechoslovak government, unlike any other government in the area, tried scrupulously to fulfill its treaty obligations to the national minorities in regard to their political representation, cultural development and economic advancement. If all the German parties were originally totally opposed to their inclusion in the Czechoslovak Republic, most of them were gradually reconciled to its existence, and from 1926 accepted a share in its government, which they retained until 1938. The Hungarian parties, continuously perturbed by the revisionist propaganda from Hungary, never joined the Czechoslovak government, but never developed a hostile attitude to the state either. As a Hungarian Prime Minister admitted to Adolf Hitler in September, 1938, they appreciated their share in the Czechoslovak land reform that the gentry-influenced government in Hungary would never match.

The Czechoslovak party system looked complex and dangerous only on the surface, and although it was criticized by foreign and domestic analysts alike, it really was a major guarantor of Czechoslovak democracy. Only a little more than a dozen parties gained enough votes at any time to obtain representation in the parliament, and none of them was large enough to rule by itself. The most that the power-minded parties could do was to combine forces with similarly oriented parties and form government coalitions, of which only a half a dozen were needed. It was, however, fortunate for Czechoslovakia that parties able and willing to collaborate in such coalitions were mostly moderate parties that commanded comfortable majorities in the parliament. The search for compromise solutions to the country's problems was often difficult and slowed down considerably the government business but it secured a fair stability in the administration of the state and a moderate course of its policies, which was acceptable to the majority of the population. It was only on two occasions, in 1920 and 1926, when a prompt agreement on government composition or programs could not be achieved, that interim governments had to be appointed by the president from among civil service experts, while the political parties continued their efforts to settle their differences. But even these "bureaucrat" governments, as they were called, enjoyed the support of the traditional coalition parties in the parliament, so that the parliamentary regime was not interrupted in Czechoslovakia even during these incidents. The most frequent participants in government coalitions were the Czech Populists (the People's Party) in the center of the spectrum, the Agrarians and the National Democrats on the right, and the Czech National Socialists and Social Democrats on the left. This was the original *Pětka* (five parties) coalition, which bore responsibility for government business from 1920 to 1926 and established the basic pattern of party rule in the First Republic for the rest of its lifetime, although some of these parties were at times replaced by others, or their number expanded, according to the outcome of parliamentary elections.

Leadership. An essential guarantee for the success of such a political system was a good rapport between an alert and enlightened electorate and responsible leaders. The often delicate compromises of the government coalitions at times required painful modifications of one's own party programs, and party members had to understand this. Only such an understanding made it possible that even parties of the right voted for such radical measures as the land reform and workers' welfare, while Socialist parties accepted the predominance of private enterprise in the Czechoslovak economy. On the other hand, the party leaders who had negotiated such compromises had to honor this trust by keeping the interests of the government and those of their constituents in balance. Their responsibility was the greater, as this system put much of the business of the government, as well as that of the parliament, in their hands. It was in the small councils of the coalition leaders that the real powers of the state rested and the fate of the Czechoslovak democracy reposed. It was a great fortune of Czechoslovakia that it had a good number of leaders who were able and willing to reason and work together and also retain the trust of the people.

The founder of the *Pětka* and the true engineer of government coalitions was Antonín Švehla, the leader of the conservative Agrarian Party, which was the largest party in the state and supplied the government with most of its prime ministers. Publicity shy, but a masterful tactician, he held the office of prime minister for most of the 1920s and designed a pattern of coalition politics that survived his retirement in 1929 and his death in 1933 until the end of the First Republic in 1938. He was loyally supported by Father Jan Šrámek of the Czech People's Party, who had coined the slogan of the *Pětka* ("We have agreed that we will agree"), and whose chief contribution to the state was that he refused to be antagonized by the violent campaign against the Catholic Church in the early postwar years, often supported by his government colleagues, and even the president, and persuaded his co-believers to rally to the republic and support its democratic regime. Socialist leaders Vlastimil Tusar, Rudolf Bechyně and Václav Klofáč rendered a similar service to the state among the workers, as they succeeded in persuading them to foresake a violent social revolution and pursue their goals gradually and by democratic means. Among the Slovaks Agrarians Milan Hodža and Vavro Šrobár, and Social Democrat Ivan Dérer had earned great merit for the rapid development of Slovakia and solid Slovak support for the government (the first as the first Slovak prime minister of Czechoslovakia in 1935–1938), although their espousal of the theory that the Slovaks had no national individuality of their own, but were — together with the Czechs — branches of one and the same nation, the Czechoslovak nation, proved to be unrealistic and even spiteful to other segments of the Slovak people. But the Czechoslovak Republic also owed much to the German leaders Franz Spina, Ludwig Czech and Erwin Zajicek, who had brought their Agrarian, Social Democratic and Christian Social parties into the government coalition and remained faithful to the state to its very end, despite unceasing and savage attacks directed against them by the Nazis.

However, the greatest among the Czechoslovak leaders was undeniably Tomáš G. Masaryk, who served the republic as its first president for seventeen years, from the nation's founding in 1918 until his retirement in 1935, he continued to influence its policies until his death in 1937. A philosopher of high moral principles and moderate inclination toward Socialism, he was wholly committed to the preservation of the state, for whose creation he had done so much that he came to be called President-Liberator. He never tired to tutor and admonish the political leaders of the country and the people at large to moderate their individual and party interests for the common good, and rarely failed to obtain the desired response. He indeed had the most unusual rapport with the people. The son of an illiterate coachman who became a university professor, a blacksmith apprentice who had turned the attention of the great of the world to his little nation and persuaded them to recognize its right to independent statehood, an academician who organized and commanded military legions to fight on all fronts in a global war for that purpose, a leader of a small party in the old empire who had become the head of the new republic; Masaryk remained faithful to his democratic convictions, his simple way of life, and constant contact with the common people. Free from every pretense, he captured a lasting admiration of much of the population of Czechoslovakia and the rest of the world. Although the constitutional powers of the Czechoslovak presidency were strictly limited, Masaryk's rapport with the people turned it into a powerful factor in Czechoslovak politics. Awed foreigners termed it the "dictatorship of respect" and identified his person so closely with the Czechoslovak regime that they often spoke of "Masaryk's Czechoslovakia," or "Masaryk's Republic." Although he was not always right and successful, and although the contributions of other leaders and groups were equally essential for the success of Czechoslovak democracy, his reputation as the greatest man in modern Czechoslovak history was richly deserved.

It was naturally difficult for Dr. Eduard Beneš, who had succeeded him in the office of the president in 1935, to live up to the example of Masaryk. He shared his philosophy and had earned much credit for the foundation of Czechoslovakia and its subsequent success as his closest collaborator in the wartime liberation movement, an eloquent spokesman for Czechoslovakia's interests at the Peace Conference, and its foreign minister for seventeen years, but he was not a charismatic figure and was less capable and successful in maintaining the influence of his office in national affairs on the same level. A member of the National Socialist Party, he was often opposed by conservative politicians and soon was challenged by an international crisis that was too formidable for any politician of a small country to confront successfully.

Foreign Relations. The foreign relations of Czechoslovakia were essentially determined by the events that had led to its foundation. Czechoslovakia was part and parcel of the international settlement established by the Paris Peace Conference and its natural place in international politics was on the side of those nations that had a vested interest in preserving the settlement, especially France. Czechoslovakia attached itself to France so closely that it often appeared to be a French satellite. French diplomats tutored Czechoslovak diplomats, French officers trained and commanded the Czechoslovak army in the first years after the war, and French economic interests often replaced those of Austria and Germany in Czechoslovak industries and banks. In 1924, Czechoslovakia and France concluded an alliance, which bound the two countries to come to each other's assistance in case of an unprovoked attack by a third country. Dr. Beneš tried to develop a similar relationship with England, but without success, because England soon turned its attention to its overseas empire and was unwilling to extend its commitments to Europe, especially Central Europe. More importantly, English interests in the larger issues of the European continent, especially the problem of Germany and Russia, often clashed with French interests, which had an unfavorable impact on British-Czechoslovak relations. The United States was favorably inclined toward Czechoslovakia from the beginning, but it soon withdrew from European politics, while Russia was forced into a similar isolation by its advocacy of global revolution.

Curiously enough, Austria — the former master of the Czechs — became the first of their neighbors with which Czechoslovakia had developed friendly relations. Anxious to prevent a reunion of Austria and Hungary under a restored rule of the Habsburgs, or a union of Austria with Germany, which would have nearly encircled the Czech lands, the Czechoslovak government even extended loans and other assistance to the inflation plagued Austrian Republic. In contrast, Hungary remained unreconciled with the existence of Czechoslovakia. It soon launched a vigorous campaign for a revision of the peace treaties that alarmed not only Czechoslovakia, but also Yugoslavia and Romania to the point that the three countries, at the Czechoslovak initiative, formed a defensive alliance against it, the so-called Little Entente, in 1921. Rather unexpectedly, Czechoslovak relations with Poland did not develop satisfactorily. Although several efforts were made from both sides to bridge over the differences concerning the small border territories, contested earlier (especially Těšín), they all failed, and the relations between the two countries remained strained until the late 1930s, when they became openly hostile. The principal reason for that was the so-called Third Europe Project designed by Foreign Minister Józef Beck to organize all nations of Eastern Europe from Finland to Greece into a bloc, under Polish leadership, which would form a barrier against a potential intrusion into the area from both Germany and Russia. Only Hungary showed some interest in this project, because it promised it some form of reunion with Slovakia and Ruthenia, which the Czechs, excluded from the project, were supposed to relinquish, but Beck persevered and made the damage to relations with Czechoslovakia irreparable.

Czechoslovak relations with Germany were first correct and neighborly, but turned hostile too after 1933, when Hitler came to power in Germany and embarked on a policy aiming at a drastic revision of the peace treaties. To increase their security, Czechoslovakia and France formed an alliance in 1935 with the Soviet Union, which felt equally threatened, although the latter was obligated to render its assistance to Czechoslovakia only if France had done so. Czechoslovakia also expended billions of dollars to modernize its army and build an extensive fortification system along its borders with Germany. Since a serious threat to Czechoslovak security could come only from Germany, Czechoslovakia felt fairly safe. But all depended on the resolve of France to resist German expansion and that had weakened tragically in the late 1930s.

The Crisis of 1938. To improve his chances with Czechoslovakia, Hitler tried to win its German minority as his ally. The two Nazi parties active there in the 1920s quickly compromised themselves in these efforts and disbanded voluntarily in 1933 to avoid a premature conflict with Czechoslovak authorities, only to reappear under the guise of a cultural organization, the Sudeten German Home Front, led by high school teacher Konrad Henlein. As such, the organization quickly gained a considerable following and in 1935, it entered the political arena again under the name of the Sudeten German Party. In the parliamentary elections of that year it won about two thirds of the German vote in Czechoslovakia.

Its success was partly due to the severe impact of the world depression, which hit Czechoslovakia in 1930 and suddenly ended the period of stabilization and expansion that had benefited the Czechoslovak economy since 1924. Although Czechoslovak agriculture survived the crisis fairly well, industrial production, always sensitive to international economic conditions, again dropped below prewar levels, causing large unemployment that reached an alarming figure of over 750,000 people, or about 16.6 per cent of the labor force in 1933. Social tensions, now fanned by both the Communists and the Nazis, caused new waves of strikes and confrontations that often assumed revolutionary fervor. But the government, equipped with new emergency powers, was able to weather the storm again by expanding its policies of social welfare and engaging in a large scale program of public works, among which defense production and construction proved particularly helpful. By 1937, aided by an increasing stabilization of the international economy, Czechoslovak industrial production again reached 136 per cent of the prewar volume and internal peace was re-established too. Sobered by the Nazi threat and Soviet alliance with Czechoslovakia, the Communist Party ceased its disruptive policies and the government hoped that it could win back the majority of the Sudeten Germans as well with a program of increased concessions that would go beyond the obligations assumed in the peace treaties.

But that was not in the book of Adolf Hitler. While England and France stood idly by, he took over Austria in March, 1938, and instructed Henlein to raise demands on the Czechoslovak government that could not be satisfied, thus creating a situation that he would resolve by taking over Czechoslovakia as well, although he was willing to bargain away its eastern provinces to Poland and Hungary, if they were willing to collaborate with him. At the same time he pretended to support the desire of the Slovak and Ruthenian autonomists for their self-government and advised Henlein to form a common front with them. Henlein obliged him in April, when he unveiled a program that would have created a Nazi German state within the boundaries of Czechoslovakia and re-oriented its foreign and domestic policies toward Germany to make it into a German satellite. Leaders of the other nationalities were prodded to announce similar demands on the government and some of them followed the German example. The government expressed its willingness to negotiate a new constitutional amendment concerning the status of the minorities, but declared a partial mobilization of the Czechoslovak army and sent troops into the border regions on May 20, when disturbances in the German districts,

combined with reported movements of German forces toward the
Czechoslovak frontier, seemed to herald a German invasion of the
country on the eve of communal elections. Hitler denied any such
movements, but secretly confirmed his intentions to invade Czechoslovakia and set a definite date for it, October 1. Alarmed by the
prospect of a war between Germany and Czechoslovakia, in which
they were unwilling to participate, England and France tried to
mediate between Prague and Berlin and sent Lord Runciman to
Czechoslovakia to attempt the same between the Czechoslovak government and Henlein's party.

But all these efforts failed. Demoralized by the sudden and easy
capture of Austria by Hitler and his massive propaganda boasting of
the German recovery under his leadership and depicting the Germans
as a superior race that should not be "oppressed" by the "inferior"
Slavs, German Agrarians and Christian Socials disbanded their
parties, which helped Henlein capture almost ninety per cent of the
German vote in the elections, and Henlein became even more intransigent in his negotiations with the Czechoslovak government. On September 12, Hitler announced publicly that his patience was exhausted
and threatened that he would resolve the Sudeten German question by
force. In agreement with France, British Prime Minister Neville
Chamberlain now took the mediation efforts into his own hands,
offering to arrange a peaceful transfer of the German border districts
of Czechoslovakia to Germany (September 15). Hitler pretended to
agree, but when the Czechoslovak government was forced to consent
to the transfer by England and France, who threatened not to support
it in case of a war (September 21), he raised his demands to include
additional districts, giving at the same time support to similar Polish
and Hungarian claims to Czechoslovak territories (September 23).
For, a few days war appeared inevitable, as Chamberlain seemed
defeated and Germany and Czechoslovakia set their armies on a war
footing. All Czech and Slovak parties, including the Fascists, rallied
behind the government, which was still backed by German Socialists
and Communists. Even the Slovak People's Party and a bloc of
Ruthenian parties, while reiterating their demands for autonomy, professed their loyalty to the common state, and while no such moves were
made in the Hungarian and Polish districts, relative calm prevailed
there during the whole crisis. Though weakened in resolve by German
influence, Yugoslavia and Romania renewed their pledges to check
Hungary, while the Soviet Union warned Poland against intervention
and declared its willingness to fulfill its treaty obligations to Czechoslovakia.

But all these pledges and assurances came to nought, when France, goaded by England, deserted its ally. At a conference in Munich, France signed an agreement with Germany, England and Italy on the night of September 29–30, which provided for the surrender of the disputed Czechoslovak districts to Germany, Hungary and Poland, more or less as Hitler dictated it. Abandoned for the second time by the western powers, the Czechoslovak government gave in again and accepted the agreement on September 30, convinced that Hitler would draw France and England into a war sooner or later anyhow, from which Czechoslovakia would rise again. These hopes were fulfilled only partially, and the First Republic remained a unique example of a small country which was able to develop and maintain a democratic regime in an area given to authoritarian traditions and preferences, in spite of all odds imposed on it by its geography and history, varied population and international pressures. As such it was viewed with admiration by foreign observers and is remembered with pride and nostalgia by those Czechs and Slovaks, and such other nationals, who had contributed to its development and strength, for they regard it as one of their crowning accomplishments in their long history.

SELECTED READINGS

Banker, Theo. *Czechoslovakia: An Economic and Financial Survey.* London, 1938. A similar survey for the first five years was edited by Josef Gruber. Financial reforms of Alpis Rašín were explained in a book of his own. Vladislav Brdlík published a survey on Czechoslovak agriculture, while A. Pavel and Lucy Textor reported on the Czechoslovak land reform. H. Böker and F. W. Büllow studied the rural exodus in the later years of the republic and Ladislav K. Feierabend evaluated the role of cooperatives in Czechoslovak agriculture. A. Kříž and V. Mareš surveyed the development of the iron and steel industry in Czechoslovakia and Alice Teichová studied its international trade.

Beneš, Eduard. *Germany and Czechoslovakia.* Prague, 1937. František Uhlíř wrote a book on the same theme. More recent monographs on Czechoslovak relations with Weimar or Nazi Germany were published by F. Gregory Campbell and Ronald Smelser.

Bruegel, Johann W. *Czechoslovakia Before Munich: The German Minority Problem.* Cambridge, 1973. Earlier works on this theme include the books of Kamil Drofta, Josef Chmelař, Elizabeth Wiskemann, Wenzel Jaksch, Radomír Luža and Bohumil Bílek. R. Gower published a book on Hungarian minorities in Central Europe and essays on the Jews appeared in a collective volume, *The Jews of Czechoslovakia*, in 1968.

Chmelař, Josef. *Political Parties in Czechoslovakia*. Prague, 1926. Karel Hoch published a similar study under the same title ten years later. Paul E. Zinner surveyed the history of the Communist Party of Czechoslovakia from 1918 to 1948.

Eubank, Keith. *Munich*. Norman, 1976. Other studies of the crisis of 1938 include books by Hubert Ripka, John W. Wheeler-Bennett, R.G.D. Laffan, F. L. Loewenheim, Henri Nogueres, Irving Werstein, Keith Robbins, Laurence Thompson, and Henderson B. Braddick.

Henderson, Alexander. *Eyewitness in Czecho-Slovakia*. London, 1939. Similar eyewitness accounts of the Munich crisis were published by G.E.R. Gedye, G. J. George, Joan and Jonathan Griffin, Duff S. Grant and Philip Paneth.

Korbel, Josef. *Twentieth Century Czechoslovakia: The Meaning of Her History*. New York, 1977. For earlier surveys see Victor S. Mamatey and Radomír Luža, eds. *A History of the Czechoslovak Republic, 1918–1948;* Princeton, 1973; Robert J. Kerner, ed. *Czechoslovakia: Twenty Years of Independence;* Berkeley, 1940; and Věra Olivová, *The Doomed Democracy: Czechoslovakia in a Disrupted Europe, 1914–1938;* London, 1972. The early years of the state were surveyed in a book by Jaroslav Císař and František Pokorný, and much relevant material can be found in Miloslav Redncígl, ed., *Czechoslovakia: Past and Present;* The Hague, 1968.

Masaryk, Tomáš G. *Masaryk on Thought and Life*. London, 1938. Other works in English reflecting Masaryk's philosophy and political views include his *The New Europe; Humanistic Ideals; The Meaning of Czech History;* and *Masaryk on Marx*. They were studied in William P. Warren's *Masaryk's Democracy;* Chapel Hill, 1941; and Antoine Van den Beld, *Humanity: The Political Philosophy of Thomas G. Masaryk*. Leiden, 1976.

———. *President Masaryk Tells His Story*. New York, 1935. Biographies of Masaryk were produced by Paul Selver, Emil Ludwig, Edward P. Neuman, Donald A. Lowrie and Victor Cohen. Zbyněk A. B. Zeman combined one with that of the president's son Jan in his *The Masaryks*. London, 1976.

Němec, Ludvík. *Church and State in Czechoslovakia*. New York, 1955. A book on the Czechoslovak Church was published by František Hník.

Opočenský, Jan. *The Collapse of the Austro-Hungarian Monarchy and the Rise of the Czechoslovak State*. Prague, 1928. For a more recent and specialized study see Dagmar Perman's *The Shaping of the Czechoslovak State: Diplomatic History of the Boundaries of Czechoslovakia, 1914–1920*. Leiden, 1962.

———, ed. *Edward Beneš: Essays and Reflections on His Sixtieth Birthday*. London, 1945. Biographies of President Beneš were produced by Pierre Crabites, Edward B. Hitchcock, Compton Mackenzie, Godfrey Lias, and

G. Granville-Geiringer. For his political philosophy see his *Democracy Today and Tomorrow.* New York, 1939.

Rechcígl, Miloslav, ed. *Czechoslovak Contribution to World Culture.* The Hague, 1964. The Czechoslovak educational system was surveyed by R. Stránský and Francis H. Stuerm.

Seton-Watson, Robert W. *The New Slovakia.* Prague, 1924. He also edited a book of essays comparing prewar and postwar Slovakia, which was also the subject of a book by C.J.C. Street. Jozef Lettrich, Jozef Kirschbaum and Jozef Mikuš wrote broader surveys of modern Slovak history. Henry P. B. Baerlein published his impressions of Slovakia in the 1920s and Ivan Dérer reported to what degree The Pittsburgh Agreement had been fulfilled by postwar developments.

Stercho, Peter George. *Diplomacy of Double Morality: Europe's Crossroads in Carpatho-Ukraine, 1919–1939.* New York, 1971. Vasyl Markus and Walter Hanak wrote books of a similar theme and C.J.C. Street published his impressions of Ruthenia in 1924.

Sturm, Rudolf. *Czechoslovakia: A Bibliographic Guide.* Washington, 1968.

Táborský, Edward. *Czechoslovak Democracy at Work.* London, 1945. Similar material can be found in Brackett Lewis, *Democracy in Czechoslovakia.* New York, 1943; and for the first ten years of the republic also in a book of Josef Borovička. The Constitution of 1920 was published and interpreted by its main author, Jiří Hoetzl.

Vondráček, Felix. *The Foreign Policy of Czechoslovakia, 1918–1935.* New York, 1937. Piotr Wandycz and Anthony T. Komjathy published monographs on French policies toward Czechoslovakia and Central Europe. Robert Machray and John O. Crane wrote on the alliance of the Little Entente, Milan Hodža on plans for a federation in Central Europe.

Wanklyn, Harriet G. *Czechoslovakia.* New York, 1954. More popular descriptions of the country can be found in the books of Edgar P. Young and J. B. Heisler.

Edward Taborsky

TRAGEDY, TRIUMPH AND TRAGEDY:
CZECHOSLOVAKIA 1938–1948

Few if any nations have gone through, within the short span of one decade, what the Czech and Slovak people experienced in the fateful period of 1938–1948. Abandoned by their Western friends in September 1938, they were forced to surrender to their worst enemy, Adolf Hitler's Nazi Germany, and to hand over large chunks of Czechoslovakia to three of their neighbors. As a result of this humiliating defeat, what used to be called "the bastion of democracy" in Central Europe was pushed into the quagmire of semi-authoritarianism. After Hitler completed the liquidation of the defenseless country by incorporating the Czech lands into his German Reich and establishing the puppet state of Slovakia in March of 1939, Czechoslovakia became the only non-German country to be occupied by Nazi Germany prior to the actual outbreak of the Second World War. Thus for almost six months the nightmare of Nazi occupation was further aggravated by the traumatic anguish that the war might come too late to save the Czech people from eventual national obliteration. Yet, when the war broke out in September, 1939, Czechoslovakia had no recognized government to speak and act in its behalf. Its leading statesman, Dr. Edward Beneš, was in exile. Beneš, whom Hitler had forced from Czechoslovakia's Presidency in October, 1939, had to wage a prolonged uphill struggle to secure recognition for the Czech government in exile.

When the long-awaited liberation finally came in 1945 after more than six years of Nazi occupation, the triumph of victory was marred by the threat of Communist and Soviet domination. Having just survived the horrors of the dictatorship of the extreme right, the Czechs and Slovaks found themselves face to face with the dire menace of the dictatorship of the extreme left. From the very outset the beleaguered Czechoslovak democrats had to defend themselves and their cause against the relentless onslaught of Soviet-backed Czechoslovak Communists bent on reshaping the country's body politic to fit the tenets of Marxism-Leninism. After nearly three years of never-ceasing confrontations the struggle culminated in February 1948 when the Communists seized power by staging a *coup d'état,* set up the coveted "dictatorship of the proletariat" and converted Czechoslovakia into a full-fledged Soviet satellite.

Munich

If one were to pinpoint the exact date on which Czechoslovakia's Calvary began, it would have to be September 19, 1938. In the early afternoon of that fateful day the British and French Ministers to Prague urgently requested a joint audience with President Beneš and presented him with the Anglo-French proposals stating in brutally clear terms that "the further maintenance within the boundaries of the Czechoslovak state of the districts mainly inhabited by Sudeten Germans cannot continue any longer without imperilling the interests of Czechoslovakia herself and of European peace" and that, therefore, areas "with over 50% of German inhabitants" must be transferred to the German Reich. In the form of a virtual ultimatum the Czechoslovak government was asked to reply "at the earliest possible moment" as the British Prime Minister, Neville Chamberlain, was scheduled to "resume conversations with Herr Hitler no later than Wednesday," i.e., on September 21, 1938.

Beneš and his government were stunned. Although they could not fail noting with growing concern the progressive weakening of the Anglo-French determination to stand up to Hitler, they did not expect that the two great Western democracies would be so blind as not to see how badly their own vital interests would be hurt and their own security undermined by what they urged Czechoslovakia to do. In particular, Beneš could not believe that France, which was committed by her Alliance Treaty to come unconditionally to Czechoslovakia's aid, would stoop so low as to pressure her ally to hand over to their common enemy the areas that contained all the major Czechoslovak fortifications. After all, as early as September 16, 1938, the French Foreign Minister sent the President (via the French Minister to Prague) a message assuring him that France would remain faithful to her commitments under the Franco-Czechoslovak treaty.

After a day-long consideration of the Anglo-French proposals Beneš and his government unanimously decided that the terms were unacceptable. Pointing out in their reply, dated September 20, 1938, that the acceptance of the proposals would sooner or later lead to Czechoslovakia's "complete domination" by Germany and to the destruction of the balance of power in Central Europe and in Europe as a whole, the.Czechoslovak government asked Britain and France to reconsider their stand and to work instead for "a quick, honorable solution" on the basis of the German-Czechoslovak Arbitration Treaty of 1936.

However, Czechoslovakia's appeal fell on deaf ears. Shortly after 2 a.m. on September 21st, a mere six hours after the Czechoslovak note had been handed over to the British and French Ministers in Prague, the two Envoys rushed back to the Hradčany Castle, the official residence of Czechoslovakia's Presidents, to tell the President that Czechoslovakia must accept the Anglo-French proposals or face "the immediate German invasion" against which she could count on help neither from France nor from Britain. Faced thus with the hopeless alternative of having to fight alone not only against Nazi Germany, but possibly also against Hungary and Poland which both were getting ready to share in the spoils, the Czechoslovak government decided after many hours of agonizing evaluation of all the relevant factors to retract its earlier rejection and to bow to the Anglo-French ultimatum. After ten more tension-packed days during which Hitler stepped up his territorial demands so much that war seemed imminent, the dismemberment of Czechoslovakia was finalized by the Munich agreement of September 30, 1938. Without bothering to invite the country whose fate was to be settled to be represented at the conference table, the chiefs of the governments of Britain, France, Nazi Germany and Fascist Italy pronounced and signed a verdict which gave Hitler a considerably larger portion of Czechoslovakia than was contemplated by the Anglo-French proposals of September 19, 1938, including some 800,000 Czechs. Nothing reveals better that the Munich agreement was really a four-Power *Diktat* than the fact that Czechoslovakia was not even asked to sign the document and the Czechoslovak government was simply directed to send its delegate to Berlin by 5 p.m. of the same day to attend the meeting of the international commission which was supposed to work out the details of the territorial transfer.

One of the major factors that Beneš and his government had to consider when confronted with the Anglo-French refusal to come to Czechoslovakia's aid was the question of what the Soviet Union would do. Under the terms of the Czechoslovak-Soviet Alliance Treaty of 1935, the USSR was obligated to help Czechoslovakia against Germany, but only if France did likewise; and, being a member of the League of Nations, the Soviet Union was also bound to help victims of aggression if and when the Council of the League made an appropriate decision to that effect. Hence, before replying to the Anglo-French proposals, Beneš felt that it was imperative for Czechoslovakia to ascertain beyond any doubt what Moscow's position would be.

Unfortunately, what he found out proved to be of little practical use in his country's predicament. The Soviet government assured him that: (1) the Soviet Union would come to Czechoslovakia's aid if France also rendered help; and (2) the Soviet Union would fulfill its obligations as stipulated by the League of Nations Covenant. But by the time Beneš was given the Soviet statement it was already known that France's help was not forthcoming and that the League of Nations was in a deep coma. After Beneš had pointed out in a talk he had with the Soviet Minister to Prague on September 21, 1938, that, in view of the French betrayal, the Soviet assurance was not enough, the Soviet envoy returned later on the same day with an improved version stating that the Soviet Union "would feel authorized to come to Czechoslovakia's aid" the moment Czechoslovakia lodged a complaint against German aggression with the League of Nations.

Beneš was grateful for the Soviet gesture which at least seemed to affirm the Soviet government's continued willingness to honor the obligations it had assumed, and even to go beyond them, and thus contrasted favorably with the faithlessness of the French and the collusiveness of the British. But it was hardly the kind of a commitment on the basis of which a responsible statesman concerned with the survival and long-term preservation of his people could issue a call to war. To begin with, the Soviet assurance given Beneš and, in fact, all the other Soviet communications and statements made during the September 1938 crisis, never specified how, when and with what forces the Soviet Union would and could actually help. All inquiries to that effect were met either with silence, equivocation or evasion. An especially grave doubt about Soviet sincerity in promising Czechoslovakia "immediate and effective aid" arose in connection with the all-important question as to how the Red Army could join the Czechoslovak Army in fighting the German *Wehrmacht* when the Soviet Union and Czechoslovakia had no common frontier. Since Poland and Rumania, the two countries that would have had to be crossed, made it clear that they would under no circumstances allow Soviet troops on their soil and the Soviet government kept insisting that the Red Army would not enter them without their permission, there was an obvious impasse. Thus when the Soviet Foreign Minister Litvinov declared in his speech in the Assembly of the League of Nations on September 21, 1938, that the Soviet Union intended to fulfill its obligation under the Czechoslovak-Soviet Treaty of Alliance "by ways and means open to us," the Czechoslovak government was bound to wonder with trepidation exactly what the Soviets had in mind. In particular, Beneš feared that the Soviet Union might have been thinking about helping only to the extent that it had helped the Spanish Republican regime against Franco. Nor could the Czechoslovak government's apprehension be alleviated when in another equivocal speech at Geneva delivered on September 23, 1938, Litvinov stated that in the event of French indifference to an attack on Czechoslovakia the Soviet Union had no obligation to help Czechoslovakia, that it "may come to the aid of Czechoslovakia only in virtue of a voluntary decision on its part or in virtue of a decision of the League of Nations," but that "no one can insist on this help as a duty." Such a way of putting it could not help but sound as a retreat from the aforementioned statement of the Soviet Minister to Prague which seemed to imply that the Soviet Union would help Czechoslovakia without even waiting for any decision by the League of Nations.

Moreover, both the President and his military advisers were convinced that Soviet Russia's military preparedness was well below that of Nazi Germany and that a war fought against the latter by Soviet Russia and Czechoslovakia alone would be lost. Also, all the information available to the Czechoslovak government at that time pointed unmistakably to the conclusion that, while willing to help if the West became involved in the war, the Soviet government definitely had made up its mind to avoid becoming unilaterally entangled in a war with Germany. As revealed by the German documents published after the War, that was also the conclusion reached by the Nazi regime. Finally, Beneš was convinced that, as the situation stood in 1938, to go to war without the support of a country other than the Soviet Union would have been to play straight into the hands of Nazi propaganda which sought to portray Czechoslovakia as a willing tool for the Bolshevization of Central Europe; and this would have provided France and Britain with an alibi of sorts for their failure to come to Czechoslovakia's aid.

The "Second Republic" and the Ides of March, 1939

Besides mutilating Czechoslovakia territorially and rendering her defenseless, Munich also sounded the death-knell of Czechoslovakia's flourishing democracy. Abandoned by the Western democracies which they admired and trusted, and forced to capitulate without firing a shot despite their elemental desire to fight, the Czechoslovak people felt humiliated and became despondent and demoralized. Such an atmosphere, coupled with the unceasing pressure exerted by the Nazi dictatorship, created favorable conditions for the growth and proliferation of authoritarian leanings.

Beneš, who continued as the President for what was left of the country, dubbed the "Second Republic," was forced to resign by Hitlerites. Beneš was replaced by a non-political figurehead in the person of the 66-year-old conservative judge, Emil Hácha, while the leader of the former conservative Agrarian party, Rudolf Beran, assumed the Premiership.

The pre-Munich Czech political parties were dissolved and replaced by two new ones: 1. the right-wing National Unity party which emerged as a dominant political force and sought to curry favor with Nazi Germany by adjusting Czechoslovakia's policies to Hitler's wishes; 2. a slightly left-wing Labor party which, due to the circumstances in which it was born, was doomed to remain weak and ineffectual. The ultra-conservative and strongly autonomist Catholic People's party acquired a virtual power monopoly in Slovakia and used its greatly enhanced political clout to secure for Slovakia (and indirectly also for Ruthenia, Czechoslovakia's easternmost province) an autonomous status with its own legislative assembly and cabinet.

However, the endeavors of the new regime to earn Hitler's good graces by conforming to his wishes proved to be of no avail. By February, 1939, the Führer decided to carry out the contingency plan, ordained just a few weeks after Munich, to "smash . . . the remainder of the Czech State." After having first induced the eager Slovak separatists to proclaim the independence of Slovakia on March 14, 1939, which he promised to guarantee, Hitler ordered German troops to invade and occupy the Czech lands on March 15, 1939. Informed of Hitler's decision during an audience with Hitler in Berlin in the early hours of March 15, 1939, and told by Goering that the German *Luftwaffe* would destroy Prague if the Czechoslovak army fought back, Hácha gave in after three hours of futile protest. He ordered that there be no resistance and signed the declaration, unprecedented in the annals of modern diplomacy, by which he agreed to "place the destiny of the Czech people and the Czech lands in full confidence in the hands of the Führer and the German Reich." As a result, Czechoslovakia was erased from the map of Europe. Its most populous western portion was incorporated into the German Reich as the "Protectorate of Bohemia-Moravia," Slovakia became a "protected state" under complete German tutelage, and Hungary was allowed by Germany to engulf Ruthenia.

Thus, for the second time in less than six months, the Czechoslovak people were made to suffer the agony of an abject surrender without a fight. Also, in less than six months, they were betrayed for the second time by their one-time Western friends. When the British and French governments urged Czechoslovakia on September 19, 1938, to transfer her German-inhabited border areas to Germany, they offered to guarantee Czechoslovakia's new boundaries. In submitting to Anglo-French pressure, the Czechoslovak government made it clear that they "accept these proposals as a whole from which the principle of a guarantee, as formulated in the note, cannot be detached." The Anglo-French commitment to that effect was also included in the Annex attached to the Munich agreement itself. Yet the commitment was not honored. In his initial reaction to Hitler's orchestrated takeover on March 15, 1939, the British Prime Minister, Neville Chamberlain, even went so far as to suggest that, since it was "internal disruption" that put an end "to the State we had proposed to guarantee," His Majesty's Government "cannot accordingly hold themselves any longer bound by this obligation." Nonetheless, two days later both Britain and France lodged formal protests with the German government refusing to recognize the legal validity of Hitler's action; and so did the governments of the United States and the Soviet Union.

The Liberation Movement Abroad

While they could do nothing for the Reich's newly acquired subjects in the "Protectorate of Bohemia-Moravia," the protests of the four Great Powers provided the much needed political and legal base on which to build and develop the Czechoslovak liberation movement abroad. It meant that, as far as France, Great Britain, the Soviet Union and the United States were concerned, Czechoslovakia, even though physically occupied by another power, continued legally to exist. Hence what was needed was to supply her with an organ that could speak and act internationally on her behalf. That was precisely what Edward Beneš, former President of Czechoslovakia and the highest-ranking and most prominent Czechoslovak statesman in exile, set out to accomplish. From Chicago, where he served as a professor at the

University of Chicago in the spring semester of 1939, Beneš sent telegrams on March 16, 1939, to President Roosevelt, the Prime Ministers of Britain and France, the Foreign Minister of the Soviet Union and the President of the Council of the League of Nations protesting Hitler's action; and he began organizing the Czechoslovak liberation movement. After the war broke out in September 1939 Beneš, who had meanwhile returned from the United States to London, initiated negotiations with Great Britain seeking to obtain recognition for a Czechoslovak government in exile without delay. Similar negotiations were conducted simultaneously with France by the Czechoslovak Minister in Paris, Štefan Osuský. However, the French and the British governments, which continued to be dominated by the men responsible for Munich, were unwilling to commit themselves unreservedly to the restoration of an independent Czechoslovakia, and were not yet ready to grant recognition to the government in exile.

After protracted negotiations they agreed in November and December 1939 to recognize only a Czechoslovak National Committee "qualified to represent the Czechoslovak peoples" (rather than Czechoslovakia as a state), and to take care of the reconstitution of the Czechoslovak army (which was being organized in France from among the Czechoslovak refugees). It was not until after the two Western signatories of the Munich agreement, Daladier and Chamberlain, had been ousted from office in March and May 1940, respectively, and after France collapsed, that further headway could be made and Beneš at long last prevailed upon Great Britain in July, 1940 to recognize the Czechoslovak government in exile. Even so, the British act was not a full *de jure* recognition, for it designated Beneš's government as "provisional" and thus relegated it to a status inferior to those enjoyed by the other exiled governments that had found refuge in England after their countries had been overrun by Hitler's armies. It took yet another year of pleading, arguing and imploring before the British finally discarded the disparaging "provisional" attribute and normalized the situation by accrediting an Envoy Extraordinary and Minister Plenipotentiary to Beneš's government in lieu of a "representative" of an undefined diplomatic status. In the same month the Czechoslovak government also succeeded, after many months of negotiations, in obtaining recognition from the United States. However, strange as it may be, while the British had already granted full *de jure* recognition, the United States came up some ten days later with only a provisional recognition; and another fourteen months passed before it followed through with a full and definitive recognition.

The main reason why Beneš was so anxious to get recognition for a Czechoslovak government in exile was to obtain an officially recognized organ for Czechoslovakia which could start working full speed for the attainment of his primary objective, namely, the complete repudiation of Munich and all its consequences. That meant, above all, the recognition of Czechoslovakia's boundaries as they had existed before Munich. But therein lay the main difficulty. To begin with, for the British it would have amounted to a public admission that they were wrong when they had forced Czechoslovakia to yield the Sudeten areas to Germany, and the assertion that "Czechoslovakia's vital interests" would be "effectively assured" by the Munich concessions would be repudiated. Furthermore, since a portion of Czechoslovak territory had been annexed by Poland in 1938, Great Britain's recognition of Czechoslovakia's pre-Munich boundaries would have pitted her against the ally in whose behalf she actually declared war on Germany. Finally, such a recognition would have run afoul of the policy, mutually agreed upon by Britain and the United States, not to assume any commitment during the war with regard to postwar boundaries in Europe.

Hence, no matter how hard they tried to make the British change their mind, Beneš and his government repeatedly ran into a brick wall. Though the British government was willing to concede that the Munich agreement, having been destroyed by Germany, was dead, it persisted nonetheless in its adamant refusal to recognize Czechoslovakia's claim to the territory that had been torn away from her by Germany, Poland and Hungary in 1938 and 1939 under its terms. Finally, after innumerable discussions and exchanges of views and memoranda, the two governments in August, 1942 settled on a compromise formula declaring "that, at the final settlement of the Czechoslovak frontiers to be reached at the end of the war, His Majesty's Government would not be influenced by any changes effected in and since 1938."

Having extracted the maximum he could get from the British, Beneš moved at once toward getting Munich repudiated by its other Western co-signer, France, whose officially recognized spokesman in exile in Britain was General deGaulle and his French National Committee. Unlike the protracted tug-of-war with the British, the Czechoslovak-French negotiations proceeded speedily and smoothly, and resulted in Beneš and his government getting exactly what they wanted. In its note dated September 29, 1942 (which happened to be the fourth anniversary of the signing of the Munich agreement), the French National

Committee solemnly declared that it considered the Munich agreements "as null and void from their inception," recognized "no territorial changes concerning Czechoslovakia which took place in 1938 and afterwards" and would do "all in its power to ensure that the Czechoslovak Republic in its frontiers of the period before September 1938 shall obtain all the effective guarantees for its military and economic security, its territorial integrity and its political unity."

However, it was obvious that, even with the best of intentions, postwar France would not be in a position to contribute meaningfully to choose "effective guarantees" that Czechoslovakia was likely to need. Moreover, considering the way Czechoslovakia was treated by her Western friends in 1938 and 1939, it is understandable why Beneš and his political associates came to the conclusion that their country's security against a new Munich must be based primarily on an alliance with the Soviet Union, which had meanwhile become Czechoslovakia's neighbor. That is why Beneš continued to maintain contact with the Soviet government through the Soviet Ambassador in London, Ivan Maisky. During the strange 1939–1941 Nazi-Soviet rapprochement, the Soviet Union virtually recognized Hitler's 1939 *coup de force* in Czechoslovakia by granting recognition to the puppet state of Slovakia and withdrawing the accreditation of the Czechoslovak envoy in Moscow in December, 1939. After Germany attacked the Soviet Union in June 1941, the Soviet government moved promptly to set things right by establishing diplomatic relations with the Czechoslovak government in exile, recognizing Czechoslovakia's pre-Munich boundaries, and concluding a new Czechoslovak-Soviet Treaty of Friendship and Alliance in 1943 which also included a Soviet commitment not to interfere with Czechoslovakia's internal affairs. A Czechoslovak army was organized in the USSR from among Czechoslovak refugees and Slovak and Ruthenian prisoners of war to fight alongside the Red Army. Upon Stalin's invitation, President Beneš paid an official visit to Moscow in December 1943 from which he returned firmly believing that he had reached a complete understanding with Stalin with regard to all that mattered. In particular, he became convinced that the Soviet regime had no intention of communizing Czechoslovakia after the war.

Before 1944, the only disappointment that the Czechoslovak government encountered in its dealing with the Soviet government was the Soviet rejection of Beneš's plan for a Czechoslovak-Polish Confederation. The sad fact that the two Slav neighbors who stood in the way of the German drive to the East were not on speaking terms with one another facilitated considerably the Nazi conquest of Czechoslovakia and Poland in 1938 and 1939. Hence, in order to prevent the repetition of such an adverse relationship in the future, Beneš sent General Wladyslaw Sikorski, the Premier of the Polish government in exile, a proposal in November 1940 for the establishment of a postwar Czechoslovak-Polish Confederation. The Polish government concurred and, after some preliminary work on the project by mixed Czechoslovak-Polish commissions, a Protocol was signed in January, 1942 providing for the creation of a postwar Czechoslovak-Polish Confederation and leaving the door open for other countries of East-Central Europe to join the Confederation.

It was at this point that the Soviets began to express doubts about the entire project, wondering whether such a policy of close cooperation with a country as unfriendly to the Soviet Union as Poland was really good for Czechoslovakia and whether such a policy was "taking a realistic view of the future." The Czechoslovak government tried repeatedly to convince the Kremlin that, predicated as it was on the establishment of friendly relations between Poland and the Soviet Union, the envisaged Confederation would be also in the best interest of the Soviet Union. But it became obvious that Stalin remained adamant in his opposition to the project, thus confronting Beneš's government with a sort of Hobson's choice: either to persist in the project and incur the hostility of the Soviets or to abandon it and anger the Poles. Since the retention of good relations with the Soviet Union was deemed to be more important for Czechoslovak security, the Czechoslovak government chose the latter alternative. Even so, Beneš continued to strive hard to salvage from his project what he could by inducing the Soviet government at least to add to the Czechoslovak-Soviet Treaty of Friendship and Alliance a Protocol allowing adhesion to the alliance of "any third state which has a common frontier with the USSR or the Czechoslovak Republic, and which in the present war has been the object of German aggression." Unfortunately, subsequent developments made the whole issue irrelevant.

In the "Protectorates"

While Beneš and his associates abroad were slowly gaining ground in their struggle to overcome the untoward legacy of Munich on the international front, their countrymen in the "Protectorate Bohemia-Moravia" and, to a lesser extent, the "Protected State" of Slovakia were being subjected to ever harsher repressive measures.

The Prague government headed by Hácha as President of the "Protectorate" was reduced to utter subservience to the German-appointed "Reich Protector" who wielded the power of absolute veto over any action of the Protectorate's government and could appoint and dismiss its members. The National Assembly and the two political parties of the "Second Republic" were dissolved and replaced by a nondescript authoritarian outfit called National Solidarity headed by a five-man National Committee appointed by Hácha. In its initial phase the exercise of the German overlordship over the Czech lands was relatively restrained. But the situation changed radically after the Czech demonstration against the German occupation staged in Prague on the anniversary of Czechoslovakia's Independence Day on October 28, 1939, and followed on November 15, 1939, by yet another demonstration by a group of Czech students protesting the death of a fellow-student who had been critically wounded while taking part in the October 28 demonstration. K. H. Frank, one of the foremost Sudeten German Czech-haters who was appointed by Hitler as Minister of State for the Protectorate, ordered the arrest of more than 1,800 Czech university students and teachers, nine of whom were summarily shot and several hundreds deported to a German concentration camp. On the same day Hitler ordered the closing of all Czech universities and colleges, thus initiating the first stage of his long-term plan to deprive the Czechs of their intelligentsia and to facilitate thereby their eventual Germanization.

Although the harsh repression never stopped, there were two instances when it reached unprecedented levels of viciousness and brutality. Such was the case in September–October 1941 when the head of the Reich Security Office, Reinhardt Heydrich, was appointed Deputy Reich Protector to act for the absentee Reich Protector, Baron von Neurath, whom Hitler considered to have been too mild. Martial law was declared, the Protectorate's Premier, Alois Eliáš, was arrested and condemned to death for having been in contact with the Czechoslovak government in exile, and some four hundred Czechs were ordered executed in the following two months. Even more drastic were the reprisals staged after Heydrich died from fatal injuries he suffered when he was attacked by two Czech parachutists in May, 1942. Thousands of hostages were taken, two Czech villages, Lidice and Ležáky, were wiped out, and more than one thousand persons were executed. It was only toward the end of the war that repression lessened as the forthcoming Nazi collapse sent Hitler's henchmen scrambling desperately for alibis or at least attenuating circumstances.

From his vantage point, Hitler had ample reason to hold his new subjects under tight coercive controls, for, save for a small bunch of Fascists and opportunists, virtually all Czechs were sworn enemies of the Reich and most of them were loyal to Beneš. As the Germans well knew, they could not trust even the members of the Protectorate's government. After all, two of them escaped in January 1940 to join Beneš's government in London. As mentioned above, the Protectorate's Premier was himself in regular contact (via the Czech underground) with Beneš. President Hácha was too, at least until 1941 when, not heeding Beneš's recurrent admonitions, he crossed the line beyond which Beneš had urged him not to go in yielding to Nazi demands and even reneged on his earlier promise to resign. In so behaving, Hácha and his collaborators honestly believed that they were serving the interests of the Czech people by clinging to their posts, claiming that they thus were able to protect them from worse abuse that would occur if real traitors and opportunists took over. They felt that the only way to minimize Nazi repression was to concede to the Germans whatever was the minimum necessary to achieve this purpose. On the other hand, Beneš and his associates were convinced that by 1940–41, the continuation in office and the concomitant collaboration with the German authorities by any Czech who had some stature and respectability prior to the Nazi takeover was harmful for Czechoslovakia's cause. Anyway, they argued, Hácha and his colleagues proved unable to avert harsh repressive measures and thus their primary argument by which they sought to justify their staying in office, especially after Heydrich unleashed his reign of terror in September 1941, was without merit and beyond excuse.

As noted earlier, virtually all of Hitler's Czech subjects were anti-Nazi and anti-German, and most of them considered Beneš to be their true leader and embraced wholeheartedly what his government stood and worked for. Furthermore, having been pro-Russian before falling under the Nazi yoke, most of them grew even more pro-Russian after the Soviet Union had been invaded by, and had stood up so bravely to, the German *Wehrmacht*. Such an attitude created a good psychological climate for an anti-German resistance movement. On the other hand, the limiting factors of geography, the woeful lack of weapons, the continued presence in the Protectorate itself of a strong pro-Nazi German minority and the ubiquitous Nazi security controls rendered effective resistance operations much more difficult than anywhere else in Nazi-occupied Europe. Moreover, the over-cautious, strictly pragmatic Czech mentality was not conducive to encouraging ventures involving great risks and inviting harsh reprisals for seemingly insignificant returns.

Hence, the various underground groups that began to operate in the Czech lands soon after their conquest by Hitler stayed away from spectacular acts of sabotage. Rather, they encouraged and engaged in acts of passive resistance, such as slowdowns in factories, delays in transportation, various boycotts, excessive red tape and spreading false rumors. But their most important contribution was their highly successful gathering of intelligence, which they transmitted via clandestine radio and frequent couriers, to their government in London. It was a veritable bonanza of valuable information of political, economic and military nature that proved to be most helpful both to Beneš's government and to its Allies.

Eventually, the Czech resistance movement did consider staging, at an appropriate moment, a general uprising. However, since neither the British nor the Soviets appeared to be willing to provide the weaponry needed for such a massive undertaking, the plan had to be abandoned. Thus, save for small-scale guerrilla-type activities in Moravia and eastern Bohemia, the only resort to large-scale armed struggle against the Germans occurred a few days before D-day in Europe when on May 5, 1945, a major uprising took place in Prague. Having seized the Prague radio station, the insurgents broadcast repeated appeals for help to the United States forces commanded by General Patton which were then only some 50 miles from Prague as well as to the Soviet forces which were farther away. Unfortunately, the Allied Commander-in-chief, General Eisenhower, chose to comply with the request of the Soviet General Antonov not to send American forces beyond the previously agreed line which left Prague in the Soviet zone of operations. Thus, by a strange twist of irony, the Prague insurgents finally prevailed over the Germans with the aid of the troops of the German-sponsored "Russian Army of Liberation" led by General Vlasov who made a last-minute about-face and turned against his German sponsors.

While the Ides of March 1939 were a day of deep national mourning, tears and desperate clenching of fists for the Czechs, it was a day of rejoicing and hopeful, though rather naive, anticipation for a good many Slovaks. For the extremist Slovak separatists, a relatively small but very active and vociferous group, the declaration of Slovak independence on March 14, 1939, was the attainment of what they had always wanted. For most of the others, it was the best, or perhaps the least undesirable, alternative obtainable given the circumstances. Even though most of the Slovaks would have rather had a fully federated Czecho-Slovak Republic of equal partnership, once it became unavailable, separate Slovak statehood was preferable to the incorporation into the German Reich that had befallen their Czech brothers and sisters or a takeover by Hungary that might have otherwise occurred. Thus the new Slovak regime led by Father Tiso began by enjoying the support of a substantial portion of the Slovak people and the acquiescence of most of the others.

However, it became evident soon enough that Hitler's notion of "independence" for Slovakia was far different from what the Slovaks had in mind. Tiso's regime had to agree to the stationing of German troops in western Slovakia and the Slovak economy was fully subordinated to German needs. Hordes of German "advisers" were supplemented in 1942 by a special Nazi "Adviser in Jewish Affairs." The "advisers" descended upon the Slovak ministries to see to it that they behaved in strict compliance with German wishes. The leader of the Nazified German minority in Slovakia was appointed State Secretary for German Affairs and thus Hitler's watchdog was planted in the Slovak government itself. The chief of the German Legation in Bratislava became a sort of Reich's governor whose "advice" was tantamount to an order. In November 1940, Slovakia was made to adhere to the German-Italian-Japanese Tripartite Act and one year later she had to join the Anti-Comintern Pact and declare war on Britain and the United States. Slovak troops had to take part in the 1939 war against Poland and in the 1941 war against the Soviet Union.

Thus, as Slovak "independence" was turning more and more into an empty shell and as the eventual defeat of Nazi Germany was coming ever closer, Slovak resistance against the Germans and Tiso's puppet regime began to increase. By Christmas 1943 the various Slovak resistance groups, including the Communists, joined together and established a common political leading organ called the Slovak National Council. In agreement and close cooperation with Beneš's government in London the Council began to prepare, as the culmination of its efforts, an uprising against the Germans in mountainous central Slovakia. When at the end of August, 1944, Hitler ordered the occupation of all of Slovakia in order to put an end to increased partisan activities, the Slovak National Council decided on August 29, 1944, to launch the planned uprising, even though it would have otherwise preferred to wait a little longer. Unfortunately, the Slovak insurgents, whose nucleus consisted of two Slovak army divisions, were not well enough equipped to be able to withstand the German counteroffensive. The aid that Stalin promised Beneš during the latter's visit to Moscow in December 1943 for precisely such an eventuality was inadequate and slow in coming. Moreover, the Soviet government vetoed the aid that might have been provided by the Western Allies, claiming that Slovakia was within the zone of Soviet military operations. Thus, after two months of desperate fighting, the uprising collapsed. Nevertheless, it went a long way toward erasing the disgrace brought about by Tiso's regime and proved beyond any doubt the Slovaks' desire to live together with the Czechs in one Republic.

The Triumph and the Tragedy

On March 11, 1945, after almost six and one-half years of exile, Beneš left England for his final journey back to Czechoslovakia — which was then in the process of being freed from the Germans by the Soviet armies. After a stop in Moscow, used to reorganize his government and to discuss matters of mutual concern with the Soviet leaders, the President and his collaborators established their first headquarters on native soil in the eastern Slovak city of Košice. A few weeks thereafter, accompanied by Madame Beneš, a small retinue of his closest associates and with a detachment of Soviet guards assigned to protect him, Beneš set out on the final leg of his journey that eventually took him back to Prague's Hradčany Castle on May 16, 1945. It was a triumphant comeback that defies description. At innumerable stops along the way from Košice to Prague the returning President was hailed as a savior by huge and ecstatic crowds expressing their gratitude, with eyes filled with joyful tears for what Beneš had done for them. The enthusiasm with which Beneš was received was a tremendous tonic he badly needed. Unlike the blissful masses welcoming him, he already knew that his erstwhile faith in Soviet good intentions was misplaced and that he was coming back under circumstances posing a grave peril for the survival of his country as a free and democratic nation.

After his return to London from his Moscow talks with Stalin and Molotov in December 1943, Beneš honestly believed that everything would be fine between Czechoslovakia and the Soviet Union and, in particular, that the Soviets were not likely to interfere unduly with his country's internal affairs. This conviction of his was by no means brought about by any sort of a starry-eyed reliance on Communist good faith. Rather, it was based primarily on his expectation that Stalin would so need Western aid in order to rebuild his country from the ravages of the War and would be so interested in maintaining Soviet Russia's newly gained status as a highly respected Great Power and amassed reservoir of good will in the world that he would resist the urge to throw wartime promises of good behavior to the winds and attempt communizing Czechoslovakia by means of force.

But Beneš's conviction that Stalin would leave Czechoslovakia alone was rudely shaken by several grave instances of Soviet misconduct in the latter part of 1944 and early 1945, the worst of which concerned Ruthenia. As soon as the Red Army pushed out the Germans and assumed control of the province, its inhabitants began to be forcibly drafted into the Red Army. Local mayors were compelled to sign "spontaneous" petitions asking that their people be "reunited with their Ukrainian brothers on the other side of the Carpathian Mountains" and similarly phrased telegrams sent to Stalin and Beneš were prominently broadcast by the Ukrainian Radio Kiev. The Czechoslovak government delegation that was supposed to take over the administration of the liberated Czechoslovak territory under the provisions of the Czechoslovak-Soviet Agreement of May 8, 1944, was denied direct radio communication with its government in London, was held virtually incommunicado and was robbed of its funds by self-styled Soviet-backed Ruthenian militia while the Soviet authorities looked the other way. When the Czechoslovak government protested these actions, the Soviet answer was that the Soviet Union could not be expected to oppose "the will of the people!" As a result, Czechoslovakia was forced to agree to cede the province to the Soviet Union even though the latter earlier had recognized Czechoslovakia's pre-Munich boundaries. As the Red Army (with Soviet security agents in tow), proceeded westward through Czechoslovakia, Soviet support was liberally given everywhere to local Communists while the activities of non-Communists were obstructed.

Thus it can be well understood why Beneš's gratification with the enthusiastic welcome with which he was received upon his return to Czechoslovakia went beyond mere personal satisfaction. Faced with the Soviet-aided process of communization, which was underway even before he set foot on native soil, and painfully aware that he could hardly count on any really effective Western aid to halt it, he knew that his only asset in the struggle to preserve democracy in his country was the trust and the support of his people and their commitment to the humanitarian legacy of his great teacher Thomas Masaryk which Beneš felt dutybound to defend. Unfortunately, as it turned out, that asset was not powerful enough to overcome the overwhelming odds with which Beneš and the forces of democracy had to contend.

Having been assigned to the zone of Soviet military operations, all of Czechoslovakia, save for a small slice of western Bohemia, fell under the control of the Red Army in 1944–45. Taking fullest advantage of the Soviet presence and all-too-willing cooperation, local Communists managed to take control over the machinery of local government, including the police, almost everywhere and use it to scare and brow-beat their democratic opponents. The composition of the new govern-ment which Beneš had to take home with him from Moscow in April 1945, was also weighted heavily in favor of Communists and their fellow-travelers. In particular, Communists managed to gain for them-selves the Ministries of Interior and Information which put them in charge of national police and the communications media. They suc-ceeded also in placing a fellow-traveling general, Ludvík Svoboda, who had been commander of the Czechoslovak army corps in the Soviet Union, at the head of the Ministry of Defense, thus securing for themselves a decisive measure of control over Czechoslovakia's armed forces. Moreover, they deftly maneuvered the Czechoslovak Ambas-sador to the Soviet Union, Zdenek Fierlinger, known for his servility to the Kremlin and the leadership of the Czechoslovak Communist Party in exile in Moscow, into the post of Premier. As a result, the longtime leader of the Party, Klement Gottwald, who was appointed as Deputy Premier, became the real chief of the government in every-thing except name.

The question perhaps may be asked at this point why Beneš and the democratic politicians allowed something like this to happen. The answer is that the way the situation had developed by March 1945, they did not have much choice. Had they persisted in declining Communist demands, negotiations would have dragged on and on while communi-zation of Czechoslovakia under the aegis of the quickly advancing Soviet army would have proceeded unchecked. As Beneš saw it, the only chance to save what could still be saved was for him to get back to Czechoslovakia as soon as possible in hopes that his presence would help and encourage the faltering democratic forces to rally and stop the growth of Communist influence. Moreover, some of the leftwing Social Democrats let themselves be lured into supporting the Com-munists in some of their demands; and so did, surprisingly and short-sightedly, the negotiators of the largest Slovak party, the Slovak Democrats, misled as they were by a Communist promise to support to the utmost the Slovak quest for the maximum degree of autonomy.

Another major advantage the Communists enjoyed was their control of the labor unions. Posing as the workers' best friend and vigorously promoting the unity of all the toilers as the main source of their strength, Communists managed to secure leading positions for themselves both at the central and factory levels of the newly organized Unified Revolutionary Trade Union Movement. Under the guise of arranging for the night protection of factories against roving saboteurs, in most enterprises they were able to set up armed squads of "workers militia" chosen from among dependable Communist workers, and thus created their own sizeable private army. Their control over the publishing and the communications media, which also comprised the authority to allocate newsprint and paper that was then in short supply, enabled Communists to provide for publicity and wide circulation of Communist and pro-Communist viewpoints while making things difficult for their democratic opponents.

In spite of all these handicaps the prospects for democracy slowly began to improve in 1946 and early 1947. Having failed by 12 percent to gain the coveted majority in the 1946 elections, the Communists began to suffer their first setbacks. Their strength in local government began to recede. They had to give up at least one ministry which was of great importance for the realization of their aims, the Ministry of Education. To the chagrin of the Party and the pleasure of the non-Communists, the Ministry of Justice, headed by one of Beneš's closest associates, Prokop Drtina, moved resolutely to prosecute police officers who placed their loyalty to the Party and their Communist superiors above their duty of impartial law enforcement. Non-Communist newspapers became more and more outspoken in their criticism of Communist malpractices. The Social Democratic party, which the Communists strove hard to cajole into cooperation with them, at long last began to shake off the hold of its fellow-traveling left wing and to denounce more strongly Communist cabals against political freedoms. The people's morale was rising and civic courage slowly began to gather momentum.

But in the summer of 1947 the situation took a sharp turn for the worse. To counter the Marshall Plan which, as he correctly figured out, was bound to weaken Communist prospects by stabilizing Europe's wartorn economy, Stalin set up a new international Communist organization, the Cominform, in September 1947, to mobilize European Communists for a resolute struggle against America's "new expansionist plans for the enslavement of Europe." An important part of this struggle was the consolidation of Communist control of Eastern Europe. Czechoslovak Communists were directed to speed up the communization process and to complete preparations for the conversion of

Czechoslovakia to a full-fledged "dictatorship of the proletariat." An additional note of urgency was injected into Communist planning when the confidential survey conducted by the Communist-controlled Ministry of Information revealed that Communists would sustain a substantial loss in the next elections. Anti-Western propaganda in the Communist-controlled communications media was stepped up and Communist attacks on the leaders of the "bourgeois" parties and their alleged betrayal of the people's interests became more vicious. In what was clearly to be the finishing touch in preparations for a coup, the Communist Minister of the Interior began removing police officers who were non-Communist from key positions and replacing them with dependable Party members. When repeated protests by democratic party leaders remained unheeded and the decision of the cabinet that the removed officers be reinstated were ignored, the Ministers of three democratic parties tendered their resignations. They expected that the President would not accept the resignations and that, unless Gottwald, who assumed the Premiership following the 1946 elections, implemented the decision, new elections would have to be called.

However, Gottwald and his colleagues had no intention whatsoever of obliging their democratic opponents by following established parliamentary practices. Instead, they set into motion their well prepared plan for the Communist take-over. They summoned their Communist cohorts to a mass demonstration which shouted approval for Gottwald's demand that the resigning Ministers be replaced by those who had "remained faithful to the people." A delegation was sent from the meeting to the Hradčany Castle to prevail upon the President to respect this "will of the people." Specially trained police detachments consisting of dependable Party members were dispatched to guard the Prague radio station, post and telegraph offices and railway stations. Regular armed forces were confined to their barracks whereas the Party's own private army, the workers' militia, armed with rifles, converged on the capital city to be ready for action in case they were needed. Communist-controlled "action committees" were set up in all ministries, agencies, organizations and enterprises throughout the country and began to purge their "bourgeois class enemies." Simultaneously, Gottwald and his associates kept pressuring the President to accept the resignations tendered by democratic Ministers and to replace them with new ones handpicked by the Communist junta. For five days Beneš, although gravely ill and virtually isolated from his democratic supporters, continued to argue, resist, and seek a compromise solution that would salvage at least a modicum of democracy for his hapless people. However, with all the odds on their side, the Communists were intransigent and Beneš finally gave in. Although he remained as President for several more months, resigning only in June 1948, he immediately left the Hradčany Castle for his private country house in southern Bohemia where death mercifully redeemed him from his torment in September of 1948.

What made Beneš capitulate to Communist demands and sign what was, as he well knew, the death warrant of the Czechoslovak democracy he had sworn to defend? The basic reason for his 1948 surrender was much the same as the one that made him submit to the Munich *Diktat* in 1938. Faced in September 1938 with the alternative either to lead his people single-handed to slaughter in a war which they were bound to lose or to avoid the bloodbath by allowing his country to be mutilated, he chose the latter course, for it seemed to him to offer better prospects for the physical survival of the Czechoslovak people. Similarly, once he came to the conclusion in February 1948 that the only alternative to surrender was a bloody civil war with what he thought was the virtual certainty of direct or indirect Soviet intervention assuring Communist victory, he opted for surrender. Beneš became persuaded by Gottwald's threats and actions that the Communist Premier would order his cohorts to fight and would use the 1943 Soviet-Czechoslovak Treaty of Friendship and Alliance to call upon the Soviet Union to "restore order." The arrival in Prague of the Soviet emissary, Valerian Zorin, further strengthened his belief that Stalin was firmly committed not to permit Czechoslovakia's Communists to be defeated. As at the time of Munich, the West was clearly unwilling to give any effective aid. Czechoslovakia's democrats, including the famed nation-wide gymnastic organization, the *Sokol,* and the legionnaires on both of whom Beneš had pinned great hopes, seemed stricken with paralysis and failed dismally to make any counter-move against the Communist-led mobs. The Communist-infiltrated armed forces were effectively neutralized. Moreover, Beneš feared that using them to disarm the Communist-controlled police and workers' militia would be sure to trigger direct Soviet military involvement. Thus, as he had done in 1938, thinking of the future rather than the present, he chose to sacrifice the happiness of the present generation to preserve intact the physical substance of his nation for the better days he was sure would come again.

Suggested Reading

Beneš, Eduard, *Memoirs; From Munich to New War and New Victory.* Boston, 1954.

Eubank, Keith, *Munich.* Norman, Oklahoma, 1963.

Friedman, Otto, *The Break-Up of Czech Democracy.* London, 1950.

Grant-Duff, Sheila, *A German Protectorate; the Czechs Under Nazi Rule.* London, 1942.

Korbel, Josef, *The Communist Subversion of Czechoslovakia, 1938–1948.* Princeton, 1959.

Mackenzie, Compton, *Dr. Beneš.* London, 1946.

Mamatey, Victor and Radomír Luža (eds.), *A History of the Czechoslovak Republic, 1918–1948.* Princeton, 1946.

Mastny, Vojtech, *The Czechs Under Nazi Rule; The Failure of National Resistance, 1939–1942.* New York, 1971.

Němec, František and Vladimír Moudrý, *The Soviet Seizure of Subcarpathian Ruthenia,* Toronto, 1955.

New Documents on the History of Munich. Praha, 1958.

Ripka, Hubert, *Munich: Before and After.* London, 1939.

———, *Czechoslovakia Enslaved: The Story of the Communist Coup d'Etat.* London, 1950.

Taborsky, Edward, *The Czechoslovak Cause.* London, 1944.

Wandycz, Piotr S., *Czechoslovak-Polish Confederation and the Great Powers, 1940–1943.* Bloomington, Indiana, 1956.

Wheeler-Bennett, John W., *Munich: Prologue to Tragedy.* New York, 1963.

Zinner, Paul E., *Communist Strategy and Tactics in Czechoslovakia: 1918–1948.* New York, 1963.

Otto Ulč

POLITICS AND JUSTICE UNDER COMMUNISM

Introduction

One of the paramount guarantees of a free people is the independence of their courts. In such a state individuals and groups, whether politically, economically or socially influential, cannot legitimately affect the course of justice. If they do interfere and manage to corrupt the judicial process — as evidenced by numerous examples, revealed and suppressed — this interference is never viewed to be lawful and proper.

In states with a ruling Communist Party the role of the judiciary is substantially different from the role in pluralistic democratic systems. For one, the principle of separation of powers and an independent judiciary is explicitly rejected. Secondly, the main function of the courts is to protect the interests of the Communist state, not those of its citizenry.

This is no calumnious accusation. The Marxist-Leninists themselves will accept such an assessment. Their philosophy of the role of the state is radically different from the understanding according to Western democratic principles. Namely, a state — any state, be that a capitalist, socialist or feudal one — is considered an instrument of violence for the ruling class. Any state — with its army, police, bureaucracy and also the judiciary — oppresses the ruled in the interest of the rulers. This is the case all over the world but with a significant difference: in the West the minority — the capitalists — rule whereas in the Communist countries the vast majority — the working people — are the rulers. Theirs is the "dictatorship of the proletariat" which, again, is no slanderous label but a proud self-assessment of the state of affairs. The dictatorship of the proletariat is hailed as the most democratic form of popular government, for it is in the service of the majority of people; unlike the case of Western democracies in which, it is charged, the state is in the service of the powerful few. The Western notion of justice, impartial and blind, is rejected as both impossible and impractical.

For example, N. V. Krylenko, Soviet Commissar of Justice, emphasized that "our judge is above all a worker in the political field," that "a club is a primitive weapon, a rifle is a more efficient one, and the most efficient is the court." Later, Krylenko himself experienced this efficiency when condemned as an enemy of the people and shot.

This subservience of the judiciary to political power has several manifestations which we shall divide into two broad groups:

First, there is the bias in philosophy and in the laws as such — "not all people are equally equal," so to speak. The law may be very specific in ordering discrimination against certain categories of citizens, e.g., a Czechoslovak governmental ordinance ordering reduction of social welfare benefits for minor orphans whose parents were labeled as enemies of the people. This is then the official bias which discriminates against a citizen outside the courtroom, prior, aside, or in addition to judicial action.

Second, there is the bias exercised by the judge who is not an independent agent and who has to perform a political role in implementing the "dictatorship of the proletariat."

The damaging impact of this dual bias has varied in time and place. For instance, far more Soviet people were killed during the rule of Stalin than under Khrushchev. In the act of ultimate paradox, Stalin managed to liquidate more Communists than Hitler did.

In the sphere of the so-called People's democracies — i.e., countries from Poland to Bulgaria — the political climate basically reflected the Soviet climate. The intensity of applied harshness varied, however. For example, it is estimated that the notorious show trials in Czechoslovakia in the late forties and early fifties rendered a toll of corpses probably exceeding the combined total in the neighboring countries. This was a tragic development in a land with such a strong democratic tradition, to which we shall now turn.

Political and Legal Order in Pre-Communist Czechoslovakia

Among the countries of Central and Eastern Europe which after World War II became part of the Soviet orbit, Czechoslovakia was unique in more than one respect.

It was the only state whose democratic form of government was not destroyed from within. In the interwar years no colonels seized power, no royal dictatorship was introduced. While the East European proletariat often lived in semifeudal conditions, with their leaders in jail or in exile, the Czechoslovak leftist movement was strong, and its representatives, including the Communists, sat in the parliament.

The country was economically developed with the non-Slovak part having achieved a high level of industrialization and decent standard of living. Furthermore, the country was only slightly scarred by the ravages of World War II, and being considered an ally by the victorious powers, no burden of reparations or permanent presence of foreign troops was imposed.

The Czechoslovak legal system reflected continental tradition, derived from Roman law and brought to modernity by Napoleonic codes. In contrast to the Anglo-Saxon common law, the continental system is based on codified rules with no regard of precedents; it is less formalistic, easier to administer, speedier, less expensive, and, in the opinion of this writer, superior overall to the system that originated in old England and was exported to her colonies, present and past. In continental countries the judges are appointed rather than elected. The security of their tenure helps to insulate them from public pressure — which a judge whose career depends upon electoral support cannot easily ignore.

When Czechoslovakia was created in 1918, it inherited two legal systems and two sets of laws: the Austrian laws in the provinces of Bohemia and Moravia, and the Hungarian laws applicable in Slovakia and Ruthenia.

A special Ministry of Information was established with the task of creating a unified legal system in the country. However, this task was not achieved by the time World War II started and the democratic republic under the presidency of Dr. Eduard Beneš was destroyed.

During the wartime years in the so-called Protectorate of Bohemia and Moravia both old Czechoslovak and German laws were in force, along with several types of courts. Their jurisdictions often overlapped. Moreover, the Nazi police machinery operated quite independently in its pursuit of terror aimed mainly at political adversaries and racial outcasts of the Third Reich.

In quasi-independent Slovakia the situation was somewhat different, partly due to the less visible display of control and implementation of Nazi interests.

Stalinism in Czechoslovakia

The defeat of Nazi Germany and the restoration of the Czechoslovak

Republic did not mean a restoration of the prewar political, economic and legal order. Only a few political parties were legalized, large industries and businesses were nationalized, land reforms were enacted and some three million Germans (the so-called Sudeten Germans, residing mainly in the border areas, i.e. each fifth inhabitant of Czechoslovakia), were expelled to Germany.

This process of expulsion evidenced the brutalization Czechoslovak society experienced in the war years at the hands of the occupiers. The concentration camp suffering and terror, especially in the aftermath of the assassination of the *Reichsprotector* Reinhard Heydrich are all too well known to require elaboration here. The tragic wartime experiences may explain, but they cannot justify the Czechoslovak postwar acts of retribution and vengeance. Retroactive laws were introduced. Kangaroo courts handed down draconic verdicts and innocent individuals were incarcerated or even executed.

In this transitional period the Communists were the most active, but not the sole political force in the country. Their monopoly of power was established by a February, 1948 *coup d'état*. Restraints and impediments to radical Bolshevik transformation of the state and society were thus removed. Among the first acts of the new regime was the promulgation of the brutal Act on the Protection of the Republic according to which real and imagined political enemies were to be prosecuted and liquidated. This in fact was the Magna Charta of Stalinist Terror administered by special tribunals called State Courts.

Among the most radical political personalities was Alexej Čepička, the son-in-law of Klement Gottwald, the first Communist president of Czechoslovakia. Čepička was put in charge of the Ministry of Justice. In this capacity he presided over the formulation and implementation of the so-called Two-Year Legal Plan. The purpose of the plan was to replace laws of bourgeois vintage with up-to-date codes. By 1951, Czechoslovakia had enacted a completely new set of codes — criminal, civil, family, etc., unified and applicable in Bohemia and Moravia as well as in Slovakia. What the democrats in the prewar republic failed to achieve in twenty years, the Communists managed in two years. It should be added though that in contrast to the slow moving machinery of the democratic state, the swift Communists moved all too often to the detriment of everybody concerned: while it had taken three generations to produce one civil code, a single generation of Communist rulers managed to produce three codes; not to mention the fact that the Communist criminal code was changed five times in less than two decades. Confusion and overall legal uncertainty were the predicted result of such energetic measures.

It was the professed goal of the architects of the new Czechoslovak legal order to emulate the Soviet codes. Alas, this model was deficient: the Soviet codes were ill-drafted, dated, and basically moribund through wide disuse by Soviet authorities.

Under the circumstances Prague found a compromise solution which can be described as "capitalist in form and socialist in content." Old inherited codes were retained in form and refurbished and infused with what was considered to be a "proper socialist spirit." Specifically this meant weakening the rules of law — thus strengthening the authority of the holders of political power. Therefore, the rules were made vague and replete with exceptions. For example, a citizen is guaranteed a right to own property, but this right may be annulled if this is in the interest of the socialist way of life; the holders of political power instruct the judge how to apply this clause of socialist propriety. In the new constitution civil rights were crippled and in fact negated by such conditional clauses and exceptions to the rule: e.g., freedom of speech is guaranteed provided this freedom is exercised to benefit the political regime.

The field of criminal law is considered politically more significant than civil law. Consequently, criminal law experienced more textual changes and more political interference than civil law. The concept of felonies was expanded both quantitatively and qualitatively. From then on, a harmless act could be interpreted as high treason. High treason, espionage, sabotage and numerous other infractions of the law carried the penalty of death. The Penal Code included over a dozen so-called verbal offences that could be committed by a slip of the tongue. An incautious political comment by a law-abiding citizen turned him into a dangerous felon.

Communist justice could hardly be performed by judges with democratic backgrounds and professional experience. It was among the first tasks of the new order to purge the legal profession — the judges, prosecutors, and attorneys — and fill the vacancies with loyal individuals of the new breed. The necessity of immediate change precluded waiting for four years until the socialist law school graduated its first class. Instead, the Ministry of Justice established a speedy course — the so-called Law School of the Toiling People — in which hand-picked Communists, mainly from among the factory workers, were turned into professional judges in no more than one year. These ill-educated political appointees then administered justice in the required

Bolshevik spirit. As political quality went up, professional quality declined.

Of all the branches of the legal profession the attorneys fared worst. They were handicapped by their image as capitalist entrepreneurs and defenders of the bourgeois order. After 1948, an absolute majority of them were prohibited from practicing law while the rest of them were incorporated into the state system of "offices of legal advice." Numerically weak, underpaid and politically insignificant, they became the Cinderellas of socialist justice. Except for serious offences, mandatory representation by a counsel in court was abolished and in civil litigation it disappeared altogether. Only the politically most trusted defense attorneys were permitted to appear in political trials where their roles were often indistinguishable from those of the prosecuting attorney. The state established a scale of legal fees so that legal representation was not expensive — but it was not effective either.

The overall picture of changes in the laws and courts appears to be a blend of barbarism and enlightenment. Long-term sentences were meted out for minor political crimes but at the same time life imprisonment was abolished with the very rational justification that such a sentence offers a prisoner no prospect for release and thus no incentive for rehabilitation. Political criteria distort the deliberations of family courts, yet the same courts administer a family code containing a number of progressive provisions on equal rights for women which to this day have not been enacted in many Western countries.

To sum up, the post-1948 legal system assisted the Communists in their transformation of the Czechoslovak state and society. The capitalists were expropriated, the propertied strata were dispossessed and the economic turnover was complemented by a social turnover. Irrespective of one's opinion about the desirability of such developments, an easy agreement might be reached as to the harshness of this process of transformation. In the prewar republic under the presidency of T. G. Masaryk, only seven individuals — all found guilty of brutal murders — were executed. In the post-1948 Czechoslovakia of 13 million people, 422 jails and concentration camps were in operation. A new generation of political prisoners was created, estimated at well over 100,000. Czechoslovakia was the first country in the postwar period known to execute a woman — Dr. Milada Horáková, former member of the Parliament and also a former inmate in a Nazi concentration camp — for a political offence. As another example, a tally of known cases shows that 6,174 monks and nuns, whose only crime was their faith, spent a total of 32,016 years in prison.

Challenge to Stalinism

Khrushchev denounced Stalin and the excesses of his rule in February, 1956. In the same year Poland and especially Hungary erupted in protest to their imposed forms of government. Yet at the same time, Czechoslovakia remained quiet, experiencing only marginal symptoms of discontent. This hibernation lasted well into the decade of the sixties, largely due to the resistance of Antonín Novotný, the president and head of the Czechoslovak Communist Party. Having been personally implicated in the crimes of Stalinism, he managed to postpone the outburst of anti-Stalinist sentiment for several years.

Finally in 1963, under the cumulative impact of domestic economic and political troubles and Khrushchev's de-Stalinization policies, the leadership of the Czechoslovak Communist Party sanctioned a debate "as to the shortcomings in the judicial sphere" in legal periodicals and also in mass media sources. The rule of law, judicial integrity, the dictatorship of the proletariat, and class struggle were among the topics.

It was a very refreshing turn of events but the debate did not go too far or too deep. The manifold criticism lacked concreteness. The elaboration of sins was not accompanied by an identification of sinners. Some victims of political trials were rehabilitated (eventually, post-mortem) but those instrumental for their fate remained untouched.

The events of 1968 presented a radically different scenario. Novotný's ruling clique was deposed and their discredited policies openly rejected. Important probes into the sordid record of Stalinist crimes were undertaken and the public was kept informed about the progress of the investigations. Matters hitherto taboo such as the mysterious death of Jan Masaryk were questioned and subjected to official inquiries. The most discredited individuals from within the judiciary were forced to resign.

The Rehabilitation Act of 1968 was a document unprecedented in Czechoslovakia or in any other country, for that matter. It subjected the entire criminal adjudication for the last two years to a review. Any person, or his relatives or heirs, could challenge the original court decision. The review court was to annul the original court decision in the event that it found violations of substantive law (e.g., false evidence), violations of procedural law (e.g., use of violence in the pretrial phase), or use of entrapment and the like. This law affected tens of thousands of citizens.

In the meantime the media of mass communication — the press, radio, and television — performed a courageous and highly constructive role in delving into the record of Communist rule in general and the brutality of past years in particular. For example, hidden evidence was unearthed, individual torturers identified and even brought before the television camera to face their former victims. A flood of testimony relating to the crimes of the political system reached the startled nation. In this process, Stalinism as a system and set of pseudo-values was thoroughly discredited and rejected.

Even the timid and conformist legal profession, in large part a product of the compromised political system, came forward in 1968 with specific remedial proposals. These included a prohibition of Communist Party interference in the work of the courts, curtailing the controlling power of the Ministry of Justice, abandonment of Soviet legal principles and practices, reintroduction of judicial independence and professionalization of the courts. Among the demands were also equal justice for all and a rejection of political favoritism and/or discrimination.

The Soviet Invasion and Subsequent Developments

The Soviet invasion of 1968 was a momentous event. It introduced a new principle in international relations known as the Brezhnev doctrine. This doctrine stipulates that the Soviet Union is entitled to intervene militarily into a fraternal socialist country that is endangered by external or internal forces, and the Soviet Union will determine whether such a danger has arisen.

Clearly, the main reason behind the 1968 invasion of Czechoslovakia was the fear of the Soviet leadership that the experiments with "socialism with a human face" posed a danger to Soviet power interests. The tanks of August crushed the ideals of the Prague Spring. However, the post-invasion political developments did not lead to a full restoration of the Stalinist practices of the fifties in politics and in the legal sphere. First, too many years had passed since the death of Stalin and the world had changed. Second, the overwhelming support of the Czechoslovak public for the Dubček regime forced its gravediggers to adopt a stand of at least tactical moderation. The most formidable impediment to the full revival of Stalinism was the Czechoslovak nation itself: the people are knowledgeable, and they were informed in 1968 about the misdeeds of the past and thus unwilling to participate in or witness the repetition of the crimes of the fifties.

Gustav Husak, who in 1969 became the head of the Czechoslovak Communist Party and in 1976, also the head of the Czechoslovak government, was himself a victim of Stalinism. He spent several years behind bars after being sentenced to life imprisonment for crimes he did not commit. Husak's inaugural statement in 1969 promised a kind of socialist *Rechtstaat*, i.e. an order in which no one will suffer because of his beliefs, and law will be applied equally to all.

This promise was not kept. True, Husak's political enemies were not hanged, but some of them were jailed merely because of their disapproval of the policies of Husak's regime. However, instances of nonjudicial persecution are far more numerous. Instead of having political adversaries arrested, they are deprived of their livelihood and so are their families. Harassment and existential persecution is the prevailing pattern.

The imprint of liberalization in 1968 was ordered to be erased. Rehabilitation of the victims of Stalinist injustices provides an apt illustration: Those who were incarcerated and later rehabilitated, are put once again on trial of "de-rehabilitation" in order to arrange that the innocent are branded guilty, for the second time.

All of the 1968 attempts to correct the errors and crimes of the past are now being viewed with great suspicion. The legitimacy of such steps is being denied. Other principles of basic decency are openly rejected. For example, Zdeněk Zuska, Lord Mayor of Prague, and formerly a prominent prosecutor, emphasized that "the principle of equality before the law has often been overused."[1] Jan Němec, Minister of Justice, castigated those judges who prefer judicial independence, including independence from the Communist party.[2]

On the other hand, not a single person guilty of Stalinist crimes was sentenced and the only trial of this kind ended in a farce. This was the case of seven officers of the secret police (including M. Pich-Tůma, a Communist member of the Parliament), who worked as executioners for the Communist Party in the style of an organized crime syndicate. In this particular case, they were accused of the murder of two men whom they had gunned down in the name of revolutionary justice and secretly buried. As a symbolic epitaph to the 1968 experiment with socialism with a human face, this only trial of the Stalinists ended in June 1969 — one month after Husák replaced Dubček as the Communist Party leader. The seven defendants who did not deny the murders as charged, were set free on both substantive and procedural grounds.

Conclusion

Since 1948 justice in Czechoslovakia followed a weird course. Stalinism was its one extreme and the emancipation in 1968 the other. Remedy of past misdeeds was prevented by an outside force. Stalinism being beyond revival, the present situation is perhaps best described as one of neo-Stalinism, with renewed emphasis on dictatorship of the proletariat, class struggle, and the subordination of law to politics. This neo-Stalinism prefers not to spill the blood of the innocent and refrains from measures of mass terror. Equally, there is nothing to indicate that this political system will reintroduce the principle of equality of citizens before the law and independence of the judiciary.

NOTES

1. *Rude pravo,* August 23, 1969.
2. *Socialisticka zakonnost,* 1970, No. 5, p. 261.

BIBLIOGRAPHY

V. Busek, ed. *Czechoslovakia.*
Czechoslovak Academy of Science. *Bibliography of Czechoslovak Legal Literature, 1945–58.*
G. Golan. *Czechoslovak Reform Movement.*
_____. *Reform Rule in Czechoslovakia.*
V. Gsovski, ed. *Legal Sources and Bibliography of Czechoslovakia.*
_____ and K. Grybowski, eds. *Government, Law and Courts in the Soviet Union and Eastern Europe.*
B. W. Jancar. *Czechoslovakia and the Absolute Monopoly of Power.*
V. Kabes. *Socialist Legality in Czechoslovakia.*

P. Korbel. *Sovietization of the Czechoslovak Judiciary.*
E. Littell, ed. *The Czech Black Book.*
A. Oxley, et. al. *Czechoslovakia: The Party and the People.*
R. A. Remington, ed. *Winter in Prague.*
Z. Suda. *The Czechoslovak Socialist Republic.*
I. Svitak. *The Czechoslovak Experiment, 1968–1969.*
T. Szulc. *Czechoslovakia.*
E. Taborský. *Communism in Czechoslovakia, 1948–1960.*
O. Ulč. *The Judge in a Communist State.*
———. *Politics in Czechoslovakia.*

George Klein

THE CZECHOSLOVAK ECONOMY

Czechoslovakia and the German Democratic Republic (East Germany) share the distinction of being the most industrialized states within the Soviet bloc. The Czechoslovak state has membership in the Council for Mutual Economic Assistance (Comecon) and the Warsaw Treaty Organization. In both these roles it makes an important contribution to the entire Soviet bloc as a supplier of sophisticated technological equipment and skills. Czechoslovakia is primarily an industrial state with only around 12 1/2 percent of its population engaged in agriculture. In 1974 agriculture contributed only 10 1/2 percent to the net material product while mining, manufacturing, and construction contributed 76 percent. Most of the industrial plant is located in the western reaches of the country, the historic Czech lands of Bohemia and Moravia. Slovakia experienced a different evolution and remained predominantly agricultural until the end of the Second World War. Since then Slovakia has undergone rapid industrial transformation and, according to government statistics, incomes between the Czech lands and Slovakia reached parity in 1971. The yawning economic gap which once divided the Czechs and the Slovaks by and large no longer exists. Nevertheless, Slovakia still lags in some ways behind the Czech lands in overall development, partially because political and economic power is wielded from Prague, the capital of the country.

Before the First World War the Czech lands formed the industrial core of the Hapsburg Monarchy. The industries of the Czech lands were the major supplier of industrial products to a far-flung empire with a population of 60 million. The industries of the region prospered in this large market protected by formidable tariff walls. The mountain chains which surround the Czech lowlands represented a natural defense network which could be used against invading forces. The

Czech lands were also traversed by a dense transportation network which is indispensable to intensive industrial development. The region is plentifully supplied with raw materials and minerals such as coal and iron ore, necessary for the development of a coal-steel economy. Czechoslovakia remains to the present day one of the most important coal producers in Europe. Its mines also supply iron ores of varying grades, and uranium, gold, silver, lead and zinc. The territories comprising the present Czech lands contained about two-thirds of the steel capacity of the Hapsburg Monarchy, as well as four-fifths of its textile industries, three-fourths of its sugar refineries, and three-fifths of its breweries. This industrial potential was wedded to a literate labor force, since the prevailing educational standards were of a high quality. Illiteracy in the Czech lands was all but non-existent — in sharp contrast to Slovakia.

The ownership of the industries in the Czech lands was to a great extent in the hands of German and German-Jewish entrepreneurs. Industrialism as a way of life came to the Czech lands from Germany and spread inland and eastward.

The creation of an independent Czechoslovakia in October of 1918 fulfilled the aspirations of the predominantly Slavic majority. It also brought with it new economic problems. The dissolution of the Hapsburg Monarchy into a plethora of successor states fragmented the internal market of the area and converted Czechoslovak industries from primary dependency on a large internal market to international exports.

David Mitrany put it dramatically:

> The results of all that dislocation have in some cases been fantastic. Goods manufactured in what is now Czechoslovak territory went before the War to the various parts of the old Empire; in the post-war years only about 10 per cent of that trade has gone to the Danubian countries, some 25 per cent to Germany, and the rest to France, England, South America, and China — though all this industry is situated many hundred miles from the nearest markets in China, India, Egypt, and South America. Austria imports onions from Egypt and tomatoes from Palestine, while Hungarian and Rumanian farmers are unable to sell theirs — again though this traffic must travel vast distances to reach the Austrian market.[1]

Czechoslovakia continued to face great problems related to economic dislocation until the inception of the Second World War. These problems were aggravated by the Great Depression, which struck in

1929. Until then, Czechoslovakia managed to balance its diverse economy by aggressive expansion into new market areas in the colonial world and in those areas which were undergoing industrialization. The Great Depression led to a worldwide collapse of international trade and Czechoslovakia, a country whose well-being was tied to international markets, was especially hard hit. The most affected were the German-speaking areas in the Sudetenland which contained many of the light industries. These tended to be the most vulnerable to a downturn in the economy because people stop purchasing luxury textiles, glassware, musical instruments, toys, etc. in periods of economic decline. The deep depression which spread over the Sudetenland found its political expression in the Sudeten German Party which gradually evolved into an appendage of the National Socialist Movement which was rising to prominence and power within Germany during the same years.

The Great Depression aggravated relationships among the several nationalities inhabiting the territory of the Czechoslovak state. Rural Slovakia remained somewhat immune to its depredations because of the agrarian structure of the society. Depressed conditions contributed to the rise of particularistic nationalism among the Germans, Slovaks, and Hungarians.

The heavy industries in the Czech lands fared relatively better because they were tied to long-term contracts under government guaranteed loans for the development of plants, railway networks, or armaments. The light industries located in the German speaking areas of the state were not cushioned from the full effects of the Depression. This deepened the political chasm which separated the Czech and German communities.

The relatively strong performance of heavy industry during the Depression exercised a powerful influence on the economic planners who charted the course for the development of Czechoslovak industry after the Second World War. Many of these were Communists or socialists who were profoundly influenced by the Soviet model. In addition, the record of the previous era tended to reinforce their bias toward heavy industry. Any investment in light or service industries was viewed as peripheral to the main thrust of Czechoslovak development. The postwar Czechoslovak state directed its main investment into new plant for heavy industrial production or the extension of output in already established industries.

The Great Depression in Czechoslovakia ended as in other Western states only under the stimulus of a rearmament drive forced upon the

country by the gathering clouds of war. The subsequent dismemberment and occupation of Czechoslovakia as a prelude to the Second
World War represents the background which forced politicians of all
shades to the conclusion that new departures would be necessary in the
post World War II era. Even the most responsible democratic representative in the wartime exile government in London wanted to draw a
line separating the politics of the prewar period. Most seemed to view
close relations with the Soviet Union as essential to the survival of a
reconstituted Czechoslovakia and were willing to realign political and
economic practice with the new realities of a situation in which the
Soviet Union would be the paramount power in East-central Europe.
The 1945 Košice Program, based on agreements reached between the
London government and the Czechoslovak Communists in exile in
Moscow, curtailed the scope of Czechoslovak democracy by limiting
the number of political parties. The Program decreed nationalization
of all major industries even before the country was fully liberated from
the German occupiers. The non-Communist parties operated under the
illusion that the Communist Parties of Czechoslovakia and the Soviet
Union would refrain from imposing their dominance. The events of
February, 1948 when the Czechoslovak Communist Party, with Soviet
support, overthrew the existing constitutional system by a domestic
coup, shattered all such myths.

The economic system which the Communists inherited in 1948 was
already nationalized. The nationalization decrees of 1945-46 affected
80 percent of all industries. Only retail trade and some service industries remained in the private sector. The Communist coup boosted the
socialist sector to an incredible 97 percent of the total economy.

The Czechoslovak Communist Party also collectivized agriculture.
The basic agricultural unit of organization was the J.Z.D. (*Jednotne
zemédélské druzstvo*), which was based on the model of Soviet collective farming. Czechoslovakia went further and faster with the col-
lectivization of agriculture than any other East European state. The
results were not rewarding. By 1962 some writers reported total agricultural production declined below the 1938 levels and the income of
farm workers amounted to only about one-half that earned by workers
in other economic sectors. The low rewards constituted a disincentive
to the agricultural population and led to the mass abandonment of
agricultural occupations. The average age of the remaining agricultural population was 50, as the most energetic and upwardly mobile
left the land. Agriculture also did not obtain the investment given to
those industries which were slated for a major buildup. Since Czecho-

slovak agriculture could not meet domestic needs, the state had to import food. This added to the burden of an already over-strained balance of payments. The collectivized system of agriculture had been the pride of the first Communist leaders, Klement Gottwald and Rudolf Slánský. The record did not justify their pride in the radical solutions they applied to the agricultural question.

No other East European state went so far in the nationalization of its economy. Most Czechs and Slovaks reacted only minimally to nationalization due to a variety of interacting factors. During the interwar period a substantial proportion of Czechoslovak industry was in the hands of elements viewed as ethnically alien. Most heavy industry and mining were controlled by German and German-Jewish interests. Many Czechs and Slovaks viewed the postwar nationalization as the means of reclaiming their national patrimony. Since the Czech middle class was essentially petit bourgeois, they defined their career aims in bureaucratic rather than business-oriented terms. Another reason for the lack of resistance was the political clout of the Communist Party which had attained a size of over two million immediately before the 1948 coup. Its inflated size resulted from a mixture of popularity, general opportunism, and overt political pressure. The huge membership of the Communist Party permitted it to avoid political compromise with countervailing forces within the society.

In the 1946 election the Communist Party polled 38 percent of the total popular vote and controlled the key ministries of Defense and Interior. From these commanding heights it could neutralize any effective opposition both before and after the coup. The split among the Social Democrats over cooperation with the Communist Party further weakened the parties which favored the continuation of parliamentary democracy.

The Czechoslovak Communist Party undertook an all-pervasive policy of leveling incomes and status. The leveling policy tended to reward those groups which manifested the greatest loyalty to the regime with little regard for the nature or quality of their economic contribution. The workers of the factory militias which played an active role in installing the Communist regime were rewarded at the expense of those with ambitions of upward mobility. Long-standing Communist Party members, and individuals who proved themselves "reliable," were rewarded with managerial positions with little regard for previous experience or educational qualifications. The gap between skilled and unskilled, professional and nonprofessional, was reduced to near insignificance. The main beneficiaries were frequently those without ambition or education. These individuals constituted a powerful vested interest opposing any restructuring of economic priorities because they felt that all reforms would be at their expense. Most economic reforms proposed a measure of incentives which would have made these groups relative losers even if overall prosperity grew. Unskilled labor and unskilled management constituted a coalition which resisted any change aimed at the restoration of performance as a criterion for rewards. This constitutes a major problem for the Czechoslovak economy to the present day. The other major brake on the development of the Czechoslovak economy, as generally agreed on by most Czechoslovak economists, was the uncritical adoption of the Soviet economic model. The Soviet five-year plans were aimed at building basic industry in a vast and underdeveloped land. A principal strategy of the five-year plan was to effect massive transfers of rural populations into urban, industrial pursuits. At the time of the 1948 Communist takeover, Czechoslovakia did not possess a vast reservoir of untapped manpower. The Czech lands had only 25 percent of their population in agriculture at the time of the Communist coup. Thus, the Czechoslovak Communists inherited a highly sophisticated industrial economy which needed fine tuning rather than a massive buildup of raw capacity. Despite this, the Stalinist model was applied uncritically to all areas of Czechoslovak life involving education, industrial production, and agriculture in the face of the reality that the Czechoslovak economy had advanced beyond Soviet achievements in most areas. In line with Soviet models, heavy industry was slated for major buildup while the already developed light industries were neglected.

Czechoslovakia has been in the throes of an acute labor shortage ever since the Second World War. This shortage resulted at least in part from the expulsion of two and a half million Sudeten Germans, which disrupted many industries in which the German labor force constituted a near-majority. For example, in January 1946, 45 percent of the workers in the glass industry were of German nationality. In the textile industry, ceramics, and paper, they constituted roughly 40 percent of the labor force. Entire agricultural areas in the Sudetenland were denuded of agricultural population. The dislocations caused by this radical population shift have not been fully healed to the present day.

The other difficulty in the initial years of post World War II industrialization was the relatively low wage rate dictated by the plan. The low cost of labor contributed to its inefficient utilization. The high reinvestment rates left the market permanently starved for consumer goods while an overabundant production of producers goods was emphasized. All of these policies led to low labor productivity in many enterprises when compared to similar branches of activity in the West. Since there was no labor market in the capitalist sense, the rewards for labor were relatively equal across occupations with little reference to individual economic contribution. In the initial years significant transfers of labor were achieved by the importation of Slovaks. As the economic gap between Slovakia and the Czech lands closed, surplus labor from Slovakia was no longer available. Increased individual productivity was the only remaining remedy. This became dramatically apparent in 1963 when Czechoslovak economic performance manifested a negative growth rate. The regime of President Antonín Novotný could not afford economic stagnation for political reasons. One of the principal promises of communism is "progress." In the name of progress, people were asked to sacrifice civil liberties and individual mobility for the collective good. This rendered economic stagnation intolerable. While the Czechoslovak standard of living might have been the highest within the bloc the Czechs did not measure their living conditions by those which prevailed in Poland, Hungary, or Rumania. Most of them were accustomed to looking to the most advanced societies of Western Europe as their model. By those standards most Czechs found the prevailing conditions wanting.

The economic conditions of the early 1960s forced Novotný's regime to turn to the academic community for solutions. The economic institutes were charged with the task of researching what had brought on the economic downturn. The economists working on this problem came to substantial agreement on the major causes. They blamed the rigidities of central planning which had led to the conditions described above. There was general agreement that greater freedom had to be granted to economic organizations and that individual initiative and productivity had to be brought in line with the rewards. They concluded that enterprises produced goods with little regard for the market because rewards were based on the fulfillment — and overfulfillment — of the plan rather than on the marketability of the goods produced. The system did not provide any market feedback and was based on the assumption that anything produced could also be sold. Planning was based on the assumption of chronic shortage, a situation which no longer applied for many products; the condition of chronic scarcity no longer held true for many goods even in the Soviet Union. The Soviets encountered similar economic difficulties by the late

1950s. Many of the products produced by Czechoslovak industries were not easily saleable either domestically or in foreign markets. Centrally planned pricing did not provide an adequate guide for demand and some inventories were glutted while other markets really did confront chronic shortages. The problems facing the Novotný regime in 1962 were not only economic but political as well. Any departure from the practices of the previous era represented a tacit admission of failure. Far-reaching changes in policy had the effect of discrediting both the regime and the system. Despite these risks, Novotný felt that there was no alternative to taking some action and began experimenting with moderate reforms by the mid 1960s. The spirit of reform spread vertically and horizontally through the whole society in ever-widening circles as intellectuals and institutions began to participate in the discussion.

The modest reforms attempted by the Novotný regime were viewed as too little, too late. Most of the groups pressing for reform wanted to introduce some form of market socialism based on the Yugoslav model. The resolute implementation of this system would have required a major reorientation of the entire economy. Such economists as Ota Šik concluded that many of the managers who owed their positions to party loyalty would not be qualified for their posts under changed conditions. Apart from the managers who possessed no qualifications, those who did came largely from narrow technical backgrounds and were production oriented. They understood very little about market forces, entrepreneurship, or incentives for their personnel. In other words, even the most capable of the managers would have encountered a new situation had the reforms prevailed.

The Yugoslav model rested on the assumption that individual enterprises produce as competitors in a socialist market and that the criteria for success are not to be measured by preplanned output levels but rather by the sales of a product under conditions of competition. Some Czechoslovak managers naturally preferred the security of the previous era in which they did not have to worry about a market. The Yugoslav model also implied wider differentiation between income levels among the populace which was anathema to groups which lacked aspirations to upward mobility. Even if it could have been proven to some of these groups that the general level of prosperity would rise as a result of the reforms and that they too would benefit, some would have still preferred the security of enforced equality to the hazards of change. These were the pressures which led to the political movement broadly described as the Prague Spring of 1968. The entire process of precipitous change remained firmly in the grasp of the reformers within the Czechoslovak Communist Party. During that period there was no discussion whether the Czechoslovak economy should be socialist or capitalist. It centered on the question as to which form of socialism was best suited to Czechoslovak circumstances. The new reforming Secretary-General, Alexander Dubček, was determined to embark on a course of wide-ranging economic experimentation.

The economic latitude available to the reformers was not as great as they would have wished. Seventy percent of Czechoslovak trade was tied to the Soviet Union and other bloc markets, much of it through long-range contracts. The reformers planned to import sophisticated Western technology in order to stimulate individual and enterprise productivity because Czechoslovak industries were hampered by obsolete equipment. In order to purchase sophisticated Western equipment, Czechoslovakia had to secure hard currency or loans. The size of the fixed obligations with the Comecon states gave the Czechoslovak economy little exportable surplus which could be diverted to Western markets. Moreover, the quality and selection of many of the items did not meet Western standards. The reformers would have faced a difficult task even without Soviet military occupation and the termination of the Czechoslovak experiment in August of 1968.

After the 1968 intervention the Soviets went to great lengths to discredit the Dubček reform regime. The Czechoslovak economy received a substantial boost from Soviet loans, some in hard currencies. The Comecon-Czechoslovak terms of trade were improved for Czechoslovak products throughout the entire bloc. This caused a tangible increase in the prosperity of the populace. The aim was to demonstrate the error of the reformers and to prove that central planning had been and continues to be the proper way to organize the Czechoslovak economy.

The problems of the present regime, headed by Gustav Husak, are more political than economic. Massive expulsion of intellectuals associated with the Prague Spring from the ranks of the Czechoslovak Communist Party deprived the society of the services of some of its best intellectuals, many of whom are now performing menial tasks. The economic configuration of Czechoslovakia is such that severe economic deprivation is not a problem for most individuals. The problem is rather the failure to fulfill the general expectations of the populace. Observers of the current scene comment on the withdrawal of wide segments of the population from political activity into consumerism.

In a society in which there are 28 cars for every 100 households, in which 86 out of 100 households own television sets, and 75 out of 100 own refrigerators, and 90 percent of all households possess washing machines, the weekend house and car are a realizable aspiration to which many devote most of their energies. One of the undeniable accomplishments of the Czechoslovak economic endeavors in the Communist era is the industrialization of Slovakia. The centralistic policies which were questionable in the highly industrialized setting of the Czech lands bore better results in Slovakia, probably because they were borrowed from a setting which resembled Slovak conditions more than those of the Czech lands.

The Czech lands were the major source of capital for the industrialization of Slovakia both before and after the Second World War. Investors obviously will have a greater voice in the use of their capital than the recipients. This equation set up the relationship of inequality between the two nationalities. Moreover, the Czech lands imported substantial numbers of Slovak workers when they were needed for Czech industrial expansion. Intermarriage and socialization resulted often in their loss of a sense of Slovak national identity. The Czech educational system made few provisions for the maintenance of a separate Slovak consciousness. Therefore the Slovak leadership favored major investments in Slovakia which would permit them to utilize their labor resources within Slovakia. These controversies were openly aired in the mid 1960s and contributed to Dubček's triumph, due to the solid backing he enjoyed within Slovakia. The emergence of a Slovak national leadership changed not only the economic parameters of the Czech-Slovak equation, but the political ones as well.

Since 1968, when the Slovaks gained de facto parity of power with the Czechs by occupying the top posts in the post-1968 apparatus, these complaints have been muted. The constitution of 1968 also decreed national parity in most ministries and economic institutions. In other words, in most ministries where a Czech holds the portfolio it is assumed that his top aide will be a Slovak, and vice versa. While there has been some erosion of this formula since 1968, few would dispute the claim that the Slovaks have gained in political influence. Alexander Dubček and Dr. Gustav Husak, both Slovaks, have been Secretary-Generals of the Czechoslovak Communist Party since 1968. Slovakia today possesses a substantially different social structure than it did at the inception of the Czechoslovak Republic. There is controversy among Slovaks if the interwar period advanced the cause of Slovak industrialization, because of the relatively weak competitive position of Slovak industries which were thrown into the same market with the more sophisticated Czech industries. Nevertheless, it remains beyond dispute that the First Republic provided Slovakia with its first effective, viable system of public education which made postwar development easier by creating a pool of literate manpower. The postwar regime, in pursuit of fulfilling its egalitarian aims, diminished the economic disparity between Slovakia and the Czech lands dramatically.

Czechoslovakia is highly dependent on the entire Soviet bloc due to the sophistication of its economy. This dependency is particularly evident in the field of petrochemicals. Czechoslovakia must import about 98 percent of its total petroleum needs. The preponderant share of this originates in the U.S.S.R. Czechoslovakia is also an importer of electric power and natural gas. This dependence on Soviet energy sources also integrates Czechoslovakia into the Soviet bloc even if there were no political pressures on its leaders. The Czechoslovak planners have responded to the dearth of indigenous non-coal energy resources by undertaking massive development in the nuclear field. The major nuclear energy stations now under construction will become operational in the mid 1980s when they will start contributing up to 30 percent of the national electricity generation. These developments remain controversial from the point of view of the environmentalists but the Czechoslovak planners feel that there are few practical alternatives. This effort is integrated with that of the other Comecon states and as a consequence, Czechoslovakia is a major exporter of nuclear technology.

Czechoslovak planners would like to increase the proportion of trade destined for Western markets, which is now about 19 percent of total trade. Czechoslovakia faces almost constant trade deficits with the West because of its dependence on Western manufacturers for the modernization of its own production facilities. As stated earlier, the only way general production levels can be boosted is by increasing individual productivity. This requires highly sophisticated technology, some of which is only obtainable in the West. Because of the long-range commitments of Czechoslovakia to the Comecon states, its planners face a formidable task in achieving even a partial reorientation of their trade. These questions are not purely economic but spill over into the problematics of Czechoslovak international relations outside the bloc area. The Czechoslovak government is faced with the dilemma of boosting consumer satisfaction, which would be facilitated by greater hard currency earnings in the West, and convincing its

Comecon partners that it remains steadfastly within the Soviet political orbit. The skill of the Czechoslovak leadership in the navigation of this perilous course will be a major determinant of its political and economic future.

NOTE

1. David Mitrany, *The Effect of the War in Southeastern Europe* (New Haven: Yale University Press, 1936), p. 173.

Ivan Svitak

THE PRAGUE SPRING REVISITED

"Injustice is with the lords, truth
with the brigand."

— From the song about Jánošík (Slovak
Robin Hood in the 17th Century)

The Experiment That Failed

The "Prague Spring" taught the Czech and Slovak peoples an unambiguous lesson about the actual role of the Soviet Union in European history. The attempt of the Czech Communist Party to adjust the Soviet type of communism to the national traditions of Czechoslovakia totally collapsed and resulted in the loss of the state's sovereignty. This event was comparable only to the defeat on the White Mountain in 1620, when the Czech protestant state was incorporated in the Hapsburg empire under the conditions dictated by the victorious Austrian Catholics. For the Western world this lesson was not new, only a demonstration of Stalinism without Stalin and another proof that the Soviet leadership was not willing to liquidate its colonial policy either in East Europe or anywhere else. The "Prague Spring" opened a promising road to the democratization of totalitarian regimes. The Czech reform Communists were modern Protestants against Stalinist papacy. They thought that the Czechs and Slovaks showed the direction in which Europe was to proceed — or that they had found a model for a new social system. That has been shown to have been only a short-lived nationalist Messianic myth.

The Czech democratization was the first phase of the revolution against bureaucratic dictatorship and the first phase of an anti-totalitarian movement; but represented only one of many alternatives and not a universally applicable model. It was the Yugoslav experiment with self-government — and not its Czech imitation — which is of key importance for European countries. The fate of the Czechoslovak state during the last thirty years, however, is a menacing actuaiity for West European states. The pro-Soviet policy in Czechoslovakia was not inaugurated by the Communists, but rather by liberal Presidents in cooperation with the coalition of democratic parties in which the Communists constituted a minority. Today the subsequent history of the Czechoslovak-Soviet friendship could serve as a source of enlightenment in the West. Unfortunately, nations and governments repeatedly confirm the veracity of Hegel's contention that the only conclusion that can be drawn from history is the fact that people refuse to learn from history.

Everybody makes mistakes. There are no infallible people or politicians, and the post-January (1968) Czech Politbureau was no exception to this rule. The old henchmen of Stalinism, cast by the circumstances into a role for which they were not fit, did not succumb to the illusions so carefully cultivated by the intellectuals. They only proved themselves to be captives of the power mechanism, party apparatus, and stereotypes of the bureaucratic dictatorship. The overwhelming majority of the nation did not share either the apparatus' outlook or the intellectuals' illusions, but realized with ordinary common sense that here was a unique chance to get rid of the bureaucratic dictatorship. For the nation, the democratization process was an equivalent of a political revolution which would remove the Stalinist henchmen from the Politbureau. The Stalinists did not head the move toward democratization, but trailed it from the beginning to the end. If anybody had illusions about the entire reform movement, it was not the general population but the representatives of the reformist line in the Communist Party, their ideological brothers at *Literani listy,* and the Western press which represented Alexander Dubček as the incarnation of national aspirations. The tragedy did not have its roots in the fact that the Czech Politbureau made schoolboyish mistakes in the estimation of the Soviets' potential actions, but rather in the fact that no independent leadership existed which could direct the radically conceived democratization.

The Soviet army intervened precisely at the moment when the alternative leadership was within reach, but did not intervene against Dubček (who failed to grasp what was going on either prior to or after the occupation, and so became the ideal liquidator of his own work). What appears as the failure of the democratization movement was really only the incapability of the Communist Party — not of the individuals, including Dubček — to lead the mass movement for socialist democracy with the postulates, personalities and apparatus of a diluted Stalinism. The "Prague Spring" was a triumph of the democratization movement, a spontaneous manifestation of the tremendous potential energy of the alliance of workers with intellectuals, a drive of the antibureaucratic and antitotalitarian opposition; in other words, a rapid sequence of events historically unprecedented in Eastern Europe. The same process was a total failure of the tendency to control such a movement by the Communist Party within the narrow institutional limits of Stalinism. The final outcome was not determined by the errors committed by the reformist Politbureau because the Politbureau itself was an error of the movement's existence.

As soon as a spontaneous mass movement gains momentum, it is no longer possible to control it from above and the developmental pace no longer depends upon an individual leader or political party. To assume that the pace is regulated by some official body is *a priori* erroneous. The most that a politician can do is to comprehend the potentialities of a given situation and modify the consequences of his actions. We can ask whether this or that politician went too far or not far enough. If we accept the question as legitimate, then it is certain that the reform Communists did not dare to go far enough, to rapidly convene the Congress, to eliminate the conservative members from the leadership, to develop workers' councils and to confront Moscow with the accomplished facts. The tempo was as slow and indecisive as the program of the January (1968) leadership which made too much noise in the mass media and too few real steps toward the creation of the mass base necessary for a radical democratization. A successful program would have been based on the following three essential conditions:

1. the restoration of civil and human rights as safeguards against police terror and autocracy;

2. the alliance of the working class with the intelligentsia instead of collaboration of the apparatuses of repression, production, and culture;

3. the promise of self-government in the production sector instead of promises for a higher standard of living.

Neither the circumstances of time nor the circumstances of place create history, mass movements, or revolutions. People make circumstances, history, revolutions and movements — people, classes, nations and races. The gross betrayal of the Czech people in 1968 by the mechanisms of alienation, by the bureaucratic dictatorship and the totalitarian state was the betrayal of the idea of socialism as an outgrowth of freedom. The Czech Politbureau tried to cover up this reality with the slogan "socialism with a human face," by the mystic thesis of a suddenly awakened "reason and consciousness." The concrete dialectics in the cadre nomenclature of the top echelons was the real name of the game. It is easy to support revolutions or radical movements so long as it is a matter of rhetoric and the personal interests of real persons are not affected. The politician's quality, however, is beyond the limit of words. Czech thought is German thought embodied in the Czech language so that Czech theoretical thought, including philosophy and politics, shares the fate of European Marxism. Every political thought lags behind practice, and so long as famous owls will fly out at nightfall, it will be so both in Prague and elsewhere. In a previous crisis, a surplus of politicizing professors and radical students shared illusions that they were *a la hauteur des principes* and anticipated history. Lenin and Marx, however, though ridiculing these illusions, would seriously consider the separation of the theory and practice of Marxism as the accomplished fact of the twentieth century. The "Prague Spring" tragically confirmed this discrepancy which, from the empirical point of view, may appear as a lack of political foresight, but in fact was the consequence of the degeneration of Marxist theory into institutionalized propaganda for the bureaucratic dictatorship.

Armed conflicts and games of chess are not lost or won by the last move because the defeat or victory ensues from the sequence of situations which precede the military occupation or checkmate. The decisive erroneous maneuver made by the reform Communists in the Politbureau in June 1968 was the resolution on the recall of General Prchlík, who was preparing an alternative to military resistance; and on the domestic front, the dissociation of the leading reformers from the *Manifesto of 2,000 Words*. Instead of convening the extraordinary congress immediately, making use of the mass support by the intelligentsia, isolating the conservatives and confronting the Moscow Politbureau unambiguously — not with the prospect of a peaceful occupation, but with a war in Central Europe — the reformists made **substantial political and military concessions that did not earn them** the confidence of the Kremlin, which broke their necks in two months. During the critical August night of the 21st, they were therefore much worse off than the Polish Politbureau in Warsaw in October 1956 and capitulated without a fight. Instead of ordering the mobilization of armed forces and defending their national independence and state sovereignty militarily and with all other means at their disposal, the reformers awaited arrest by the occupation armies. They were potentially — and actually — guilty of high treason, and they acted as collaborationists and criminals according to the most elementary rules of state and international law.

What are armies for, if they do not defend a country against aggression? What is a nation good for, if it gives up its freedom? What is the state for, if it surrenders its sovereignty? In August, 1968, the Czechs should have fought even under desperately unfavorable conditions because the nations, states and institutions which surrender their highest values without a fight eventually inflict upon themselves greater damage than military defeats and loss of life. Although the military resistance could not have lasted longer than several days and the consequences would have been similar to those in Hungary, the cruel malevolence of history could have, paradoxically, created a more favorable result than the bloodless capitulation has. According to the old Greek fable, warrior Brasidas once caught a mouse which bit him. Brasidas let the mouse go and said: "Whatever is capable of defending itself deserves to live." Brezhnev is certainly no Brasidas, but only states, nations, institutions and individuals incapable of living do not defend themselves when their very existence is at stake. In the light of the history of Finland, Romania, Yugoslavia, Hungary, or Israel, there was no excuse for the 1968 Czechoslovak capitulation.

The surface of events is always deceptive because the situations and phenomena are not determined by themselves, but in conjunction with other phenomena and events. The disaster of occupation did not signify the end of the democratization movement, but, on the contrary, produced the highest degree of national unity: the overwhelming majority of workers and intellectuals united against the military occupation apparatus and waited for the decisive steps of the leaders of democratization. In the fall of 1968 this situation made the role of the reformist bureaucrats particularly deplorable because, apart from announcing their absolute failure on television, they did not manage to utilize the potential of national unity for political purposes. In order to stay in office at any price, they prevented any mass actions or at least systematically stabbed them in the back, imposed martial law against their own followers, restored censorship and then conscientiously worked on the disintegration of that movement which had lifted them up from the anonymity of history to the European stage and then threw them back into historical obscurity.

Apart from military resistance, it was still possible to defend the democratization economically, on the international forum, in the United Nations, at the conference of Communist parties and by assuming risks which would have made things uncomfortable for the Soviets. Instead, the reformists went to their defeat in a disciplined manner and did not even try to sell their own skins for a high price. The history of the Czech Communist Party is a history of betrayals of the ideals of humanism and socialism, but the phase between August, 1968 and April, 1969 was definitely the most disgusting and frightening period because the apparatus of the bureaucratic dictatorship systematically broke the back of the nation and the necks of the best people that modern Czechoslovak history had produced. The firing of salvos by Czech police into the demonstrating workers on the first anniversary of the occupation (August, 1969), was the climax, the opening of the road toward the "normalization" which revealed the fully re-Stalinized institution of the Czech Communist Party in its inglorious rule. The shots fired into the crowd, police terror, censorship and the silence of the graveyard — this indeed is the road to the normal state of things under the bureaucratic dictatorship because the system must suppress human and national rights for the sake of its own existence. During the occupation and after, the reform leaders Dubček, Smrkovský, Svoboda, and Husák acted in a way similar to the men at Munich in 1938. The only difference is that today the spirit of Munich bears another name — that of normalization of relations, or detente.

Normalization is the restoration of the bureaucratic dictatorship on Stalinist economic, political, and cultural foundations; and if the bureaucratic dictatorship is the measure of normalcy, then the return of the Novotný regime without Novotný is a true consolidation process. On the other hand, however, if we regard a measure of human freedom as the criterion for social systems, then the return to the abnormality of terroristic practices of the dictatorship represents a further stage of disintegration of Stalin's legacy and is the normalization of the abnormal. Normalization is the logical consequence of the abortive experience with democratization: Stalinist structures cannot be made more democratic, either by fair means or foul, and the bureaucratic elite must, at the decisive moment, suppress its own will to reform because this would threaten its own existence. Normalization signifies the humiliating acceptance of semi-colonial status by a bureaucratic clique of collaborationists who, in contrast to the previous clique of *apparatchiks,* lost their mass base and have thus become directly dependent upon the Kremlin. In this condition as a socialist protectorate, the population acts prudently, if it accepts the defeat and tries to live in the occupied country as well as is possible under the circumstances. That is no more mysterious than the behavior of Frenchmen in the years 1940–1944. The myths of resistance are created *ex post facto*. For a mouse in a trap, eating the bait is the only alternative left.

Totalitarian dictatorships do not collapse even in extreme situations, if the mechanisms of terror, police and surveillance preserve the cohesion of leading terrorists. Bureaucratic dictatorship does not depend on internal support, but upon the continuity of fear which makes organized actions against it impossible. In addition, Husák's regime has its internal support in approximately 100,000 stratum of the bureaucratic elite which are directly interested in the continuity of the political system. If military dictatorships can hold out for decades with even a narrower base of support, then there is no reason why the Czech bureaucratic dictatorship could not survive the test of history. The Soviet bureaucratic dictatorship endured the most difficult conditions, yet it has always been able to crush any resistance. Husák's regime does not rely on the Soviet army and the Kremlin alone, but it is highly improbable that it will hold out without substantial modifications if the Soviet armed forces were to withdraw from Central Europe. Husák, of course, is such a skillful politician that he would be willing to complete the unfinished part of Dubček's role. Stalin did not feel ashamed about implementing the political program he had stolen from his own sworn enemies, the Trotskyites. Yet, according to a well-known dictum, history repeats itself only as a farce of previous tragedies and it is doubtful that in the next ten years the young and middle generation in Czechoslovakia will fall for any pacifier of democratization under the banner of a sclerotically senile institution like the Czech Communist Party. Common sense remains a natural antidote to the propaganda of fraud.

The Future of Bureaucratic Dictatorship

> "Just imagine Austria disintegrated
> into several republics. What an
> opportunity for Russian czarist
> imperialism!"
> — František Palacký

Czechoslovakia's future in the twentieth century will be fundamentally affected not only by the internal problems of the bureaucratic dictatorship, but by European politics as well. This is the fate of small nations, although one can say that even today's "big" European nations are plagued by the superpowers to a much larger extent than they themselves wish. Czechoslovakia can exist only as a part of supranational formations, and whether it will be formally independent and autonomous during the next few decades is of little importance for the formation of suprastate entities. Under these prospects, only three eventualities are possible:

1. Czechoslovakia will be a part of the Soviet bloc, possibly of the Soviet Union itself, which in fact represents the continuity of the present situation and the most probable variant, but is not nearly as hopeless as it may appear at first glance, because changes within the Soviet system are inevitable.

2. Czechoslovakia will become a part of the federalized Central Europe or, to put it more accurately, of the atom-free zone which would serve as a buffer zone between West and East, and would constitute part of a "Finlandinzed" zone. This does not seem realistic at the moment, but it is by no means impossible if the conflict between China and the USSR intensifies.

3. Czechoslovakia will join the United States of Europe which will come into being as a superstructure of Western Europe, after their economic integration and as an attempt to save the subcontinent from European nationalism — which the Kremlin exploits with admirable skill.

Eventualities No. 2 and 3 are only possible as a purposeful defense of Western Europe against Soviet imperialism or in case of a turmoil in the Soviet Union. Modern history manifests such a strong sense for the implementation of paradoxes that precisely the most extreme eventualities are the most probable. A realistic politician expects the impossible, anticipates absurdity as the rationality of tomorrow, and accepts the unimaginable as the crucial factor in future history.

The "Moscow Spring" depends upon detente. Detente is a new name for that phase of the Cold War which ends by the normalization of the abnormal situation after World War II, without touching upon the key problems which preceded the Second World War. The hopes that the Soviet Union will relax its internal policies, however, rapidly vanish so that the "Moscow Spring" is not in sight. More likely prospects are a generals' *coup d'etat,* a war with China or the disintegration of the Soviet Union into a number of national states. Naturally, we also cannot rule out a new Khrushchev and further controlled liberalization following Brezhnev's neo-Stalinist phase. Alas, foreign policy in regard to East Europe would remain unchanged even after a change in the Soviet leadership so that the Czechoslovak future would be as gloomy as before, should we pin our hopes on the benevolence of the new, post-Brezhnev Soviet leadership. It is worth remembering that Stalin did not occupy Yugoslavia, while Khrushchev organized the Hungarian massacre. It is difficult to ignore Karl Marx's authoritative opinion on Russian policy toward Europe: "First of all, the Russian policy is unalterable as admitted by the official historian, Moscowite Karamzin. The Russian methods, tactics, maneuvers can change, but the Polar Star of Russian policy — the domination of the world — is the fixed star."

Like South American cliques of generals and dictatorships which are not affected by government changes in Washington, precisely because the mutual interests of the countries involved are permanent, so a governmental change in the Kremlin would not necessarily signify an improvement for the occupied countries of Eastern Europe. If Russian imperialism survived the revolution and Lenin, it will also survive men of smaller caliber who head and will head the Soviet state establishment and czarist imperialism in the revolutionary wrapper.

But let us ask an academic question: What would happen if the Russians left? First of all, Gustáv Husák, that most skillful politician of the "Prague Spring," would borrow Dubček's sweet-toned lyre. He would try to play out the role interrupted by the occupation — a role which, in my opinion, he played very well, much better than Dubček with Smrkovský. At the same time, the fundamental problems of democratization would reach the same point within a few months as in July, 1968 — that is to say a point of crisis. The nation, skeptical of any new fit of enthusiasm, would accept very little as a substantial improvement. After all, the Husák-Dubček coalition already existed once; why should it not be possible to sell it to the nation again, in a caricatured and equally fraudulent edition, as part of the clearance sale of the principles of socialism? Although it is hard to believe, the return of Novotnýan tolerance of the 1960s would not only be presented, but also accepted as an unquestionable step forward which in fact it would be — after an era of occupation. The reform Communists, who suffered some discrimination, would again find themselves in the spotlight of history and the decree permitting the payment of royalties in hard currency would be a sufficient ideological argument for writers to join a literary club.

Dubček as well as Husák are the representatives of the bureaucratic dictatorship; both of them are men of the institution (i.e., the Party), which is the backbone of the entire system. In a situation critical for the preservation of the regime they would do everything in their power to save the system. The normalization phase which enabled the consolidation of the state apparatus was sufficiently long to make the bureaucracy feel strong enough to avert potential turmoil through minimal concessions. But it is precisely here that the bureaucracy will be confronted with the impossibility of resolving the fundamental economic questions present within a bureaucratic system of government. And the explosive situation inaugurated in January 1968 is here again, two thousand tanks are waiting for two thousand words. The Polar Star of Russian imperialism is immutable. The theories may have many various meanings, but a slap in the face has only one. They were, however, taught an unforgettable lesson which their geographical position and Great Russian chauvinistic imperialism will never let them forget.

The idea of humanistic socialism existed in Czech history at least fifty years prior to the foundation of the Communist Party and will exist at least fifty years after the demise of this institutionalized caricature of Marx's ideas. The natural sclerosis of the Party adds to its comicality all the more when this institution tries to look revolutionary. Socialist democracy is for the Czechs and Slovaks as vitally urgent as for Western Europe, because the program of freedom and reason is the original objective of both. Ideologies, these modern substitutes for religions and secularized philosophy, have a vigorous life, and because they are not rational, history does not do them much harm. The philosophies and political ideologies derived from Marxism are vestiges of the last century, the specters frequenting only plush salons. In the battle of political ideologies the rational man thus stands outside the combat because ideological duels are only verbal skirmishes of group, class and race interests. The real problems, however, which remain for science, art and politics after the organized fits of enthusiasm and/or disgust are immense and call for that transformation of social structures which Marx anticipated so accurately.

The future political system of Czechoslovakia casts us into the sphere of "speculative history" or into an area of uncertainty. First of all, what was vaguely called democratization (a sort of liberalization of the totalitarian bureaucratic system), was a historical necessity and a larger measure of civil rights, relaxation of police pressure and some decentralization will be carried out no matter what happens. In other words, the modification of the bureaucratic dictatorship will take place anyway. The question, however, is whether this modification will

be effected from the right or the left, or to put it differently, what kind
of democratization — with violence or in a peaceful way, gradually or
suddenly — will disrupt the bureaucratic dictatorship? What the
German revolution of 1848 had not achieved, Bismarck did, and what
Dubček promised can be honored by Husák even without realizing the
contradictory nature of this development. The possibility of transition
from bureaucratic dictatorship to socialist democracy is the new iron
law of history. Ananké does not stop in front of the Iron Curtain and
create a unique dialectic of the relationship between the popular
socialist demands and the power elite, between the bureaucratic dictatorship and the Communist Party. History, the lover of paradoxes,
inaugurates this process by the disintegration of the general staff of the
dictatorship, while the institution of the Party could be used under the
indifferent anonymity of history for the disintegration of the bureaucratic dictatorship itself. This, however, is only the beginning of the
game which was played with classical seriousness during the Prague
Spring; after the explosion and collision, the dramatic sequence was
interrupted by the tank divisions. Ananké has enough time for subsequent installments.

The system of bureaucratic dictatorship can be replaced only by a
socialist revolution — a profound structural change in consciousness
and organizational forms of social life. This change must be in any
event the result of peoples' activity, a mass movement. The democratic
revolution, not necessarily a bloody one, is as indispensable a phase of
the historic process as were the economic changes after World War II.
The system of bureaucratic dictatorship cannot be destroyed without
mass support and any orientation which is satisfied with the half-
hearted measures of Dubček's leadership will also inherit the consequences. Without the democratic revolution, there will be neither
socialism nor freedom in Czechoslovakia although the political system
can be made more acceptable than it is today. Freedom is the realistic
possibility of action, not only a notion; it is the reality of political
commitment of the nations, peoples, and classes of history, not the
ideology of "reason and conscience" or the vagueness of "socialism
with a human face."

Socialist democracy is a broader term than self-government of the
Yugoslav type, but even in today's form it would represent a considerable improvement and perhaps a working system which would
finally combine economic democracy, based on the collective ownership of the means of production, with the only permanently progressive, imperfect and vulnerable, but nevertheless best form of adminis-

tration of general interests — democracy. The assumption that such a combination is possible has so far had little support in history — it is still an unfulfilled vision. Every form of socialism is based on the postulate, shared with anarchists and utopians of various shades, that economic and social cooperation of people is possible without outside compulsion. Is it true? And how is the problem of the diversity of human interests to be solved, if the human community is not to sink into the uniformity and conformism of egalitarianism? The only non-utopian solution is such a concept of human society in which the human diversity of the sovereignty of the individual (human freedom) is accepted and the sphere of compulsion (economic activity and the activity of the state) is recognized, because some norms of behavior must be imposed and enforced by any society. Such a social organization is the socialist democracy, the self-government of production and freedom of culture; in other words, a non-anarchistic and non-utopian alternative to the bureaucratic dictatorship with a state-controlled economy and guardianship over cultural affairs. In the ČSSR and USSR the bureaucratic model of constructing "socialism" served well until its 50-year existence in the USSR and 20-year existence in the ČSSR confirmed that socialism cannot develop from these bureaucratic systems of state capitalism because it was never inherent in them as a developmental potentiality.

In the future, Czechoslovakia can exist as an autonomous state only in the federalized United States of Europe, but not otherwise. The independence of Czechoslovakia was questionable both during the interwar period, when Czechoslovakia depended on France, and during the postwar period, when it depended on the Soviet Union. The history of the Czech nation makes it too clear that the Czechs cannot defy the determinant of geography and are always defeated when they cast themselves into a Messianic role and overestimate their strength. The eventuality of the "Finlandization" of Europe ceases to be a phantom of the future and appears as an acceptable alternative. The United States of Europe, which is the only exclusive alternative to "Finlandization," is one of the few hopes which a skeptical mind entertains. For the time being, however, it seems that if the West European states continue along the present course, the decision on the incorporation of the Czechoslovak state into such a federation of European states will be made in the Kremlin. The question would then be whether the USSR would be willing to make political concessions toward the liberalization of its system, which seems to be improbable.

Jan F. Triska

CZECHOSLOVAKIA: THE POLITICS OF DEPENDENCY*

I am forced to conclude that the present party line has been "successful" in the sense that social values have been successfully destroyed, but that none of the conflicts which led to the crisis in the party and society in the 1960's [before 1968] have been solved.

So wrote Alexander Dubcek, the eclipsed architect of the Prague Spring, in a letter to the Czechoslovak Federal Assembly reportedly penned on October 28, 1974, the 56th anniversary of the birth of Czechoslovakia in 1918. Dubcek's long, critical statement was later smuggled out of Czechoslovakia, allegedly by members of the socialist opposition, and received worldwide press coverage in mid-April 1975.[1] The same week saw publication in the West of an open letter from the noted playwright Vaclav Havel to Gustav Husak, Secretary General of the Communist Party of Czechoslovakia (KSC) and President of the Republic, protesting against the pervasive system of "rule by fear" in the country.[2] At the end of April, Ludvik Vaculik, a prominent novelist and writer, sent a letter to Western news agencies in Prague complaining of police harassment against himself and other well-known individuals.[3] And at the end of May, Karel Kosik, the eminent Marxist philosopher, reported similar acts of harrowing intimidation in a letter to his French colleague, Jean-Paul Sartre.[4]

These messages are just a few links in a long chain of denunciatory letters, interviews, statements, declarations, and documents that have been sent out of present-day Czechoslovakia and published abroad. Despite the outwardly quiet, orderly, and even somewhat prosperous facade of the country ten years after the Warsaw Pact invasion, the outward flow of critical and often poignant commentary suggests the existence of deep tensions in Czechoslovak society. The fact that the complainants, who include many ordinary citizens as well as known figures, have been willing to risk further retaliation by allowing — and in some cases actively seeking — publication of their protests in the West is an indication of how serious they themselves view the situation to be.

A Voice from the Grave

Prior to the spate of protest letters (which will be discussed in greater detail below), perhaps the most dramatic and revealing statements to emerge from Czechoslovakia were several messages issued by Dubcek's former colleague, the late Josef Smrkovsky who — as a member of the KSC Presidium and Chairman of the National Assembly in 1968 — was one of the four most prominent personalities of the Prague Spring (together with Dubcek, Premier Oldrich Cernik, and National Front Chairman Frantisek Kriegel). Smrkovsky's statements were made in the years 1971 through 1973, when he was engaged in a losing struggle with bone cancer, to which he finally succumbed on January 15, 1974. Possibly his illness increased his determination to speak out with candor about past and present events; certainly, in any case, he issued some extraordinary and eloquent appeals addressed to a solution of the ills brought on by the occupation of Czechoslovakia. Some of these messages will be examined in a later context; of particular interest at this juncture are two taped interviews which were sent to and eventually made public by Davide Lajolo, Editor-in-Chief of the Italian weekly, *Giorni — Vie Nuove,* and a member of the Central Committee of the Italian Communist Party. Since the more recent of these two interviews deals with events dating back to the preinvasion era, it seems useful to discuss it first.

According to the published account, the interview was conducted in a hospital during the late stages of Smrkovsky's illness. In initiating a serialization of the interview on February 20, 1975, Lajolo announced in a foreword that he had received the tape before Smrkovsky's death but had complied with the Czech leader's wish to delay publication for one year.[5] Avowing that "freedom cannot be separated from socialism — it is precious and inalienable to us Communists," Lajolo defined his purpose in publishing the account as follows:

> We have never accepted the position that one socialist country should have on its territory armed forces of other socialist countries to guard it or guarantee it. . . . This publication is meant as a service to the countries and parties of the Warsaw Pact . . . *to examine the question of the occupation anew.* We know that *consultations on this subject*

have taken place already; we know also that internal resistance in Czechoslovakia is stronger than in the other countries of the Warsaw Pact. . . . Is it just that so many Communists should live like Dubcek? Is it just that Smrkovsky had to die without the card of a party he had served his whole life?[6]

The substance of the interview provides new information on the behind-the-scenes events of the short-lived Dubcek era, as well as on the dissensions and uncertainties that plagued the reform Communists from the start. The account covered the period from the October 1967 plenum of the KSC Central Committee, which initiated the fall of the late and unlamented party boss, Antonin Novotny, to the post-invasion meeting of the Soviet leadership with those of the Czechoslovak leaders who were forcibly flown to Moscow in the last days of August 1968.

According to Smrkovsky, discussions in the Presidium in December 1967–January 1968 produced less disagreement among the top party members about forcing Novotny out than about who should succeed him. Dubcek's name at first evoked little enthusiasm, and he was in fact reluctant to accept the job. But Dubcek, then the party chief in Slovakia, was the only person against whom there was no real opposition from either the Slovaks or the Czechs. Once assured of the general support of all factions, he acceded to pressure and assumed the leadership.

Smrkovsky furnished many hints of the new regime's troubles with Moscow as well as with anti-reform elements in the Communist Party. For example, on May 6, 1968, Dubcek, Smrkovsky, Cernik, and Vasil Bilak, the ultra-conservative Slovak First Party Secretary, met with Soviet leaders in the Kremlin to discuss the Czechoslovak economic situation. According to Smrkovsky, the Prague delegation spent the day listening to Leonid Brezhnev's blistering criticism of recent developments in Czechoslovakia. Brezhnev supported his detailed attack with voluminous clippings from Czechoslovak newspapers and with other materials and reports evidently furnished by anti-Dubcek forces, some never before seen by Dubcek and his friends. In the ensuing discussion Bilak took the Soviet side. Discouraged by the bitter Soviet reproofs and weakened by Bilak's behavior, Dubcek, Smrkovsky and Cernik did not fare well in their mission to obtain Soviet help. When they asked for a loan of half a billion rubles, Aleksei Kosygin and others reiterated a position which, in Smrkovsky's view, was designed to limit Czechoslovakia's independence from the Soviet Union.

Arguing that no one in the West or East was interested in Czechoslovak consumer products but that the socialist camp wanted Czechoslovak capital goods, the Soviet leaders apparently insisted on the need
to continue emphasizing heavy industry (the old position of the "iron-
eaters"), a policy which had made Czechoslovakia dependent on the
USSR for both raw materials and markets and rendered its economy
inefficient in the past.

Smrkovsky revealed some heretofore unknown details about the
crescendoing confrontation between Prague and Moscow. For example, in the meeting at Cierna-nad-Tisou on July 31, 1968, Soviet
Politburo member Piotr Shelest confronted the Prague delegation
with an accusation that leaflets printed in Czechoslovakia were being
circulated in Ruthenia — formerly part of Czechoslovakia but now in
the Soviet Ukraine (UkSSR), for which Shelest was immediately
responsible — asking the population to secede from the USSR and
join Czechoslovakia. According to Smrkovsky, the Czechoslovak
team considered the accusation so baseless and provocative that it
declared further negotiations impossible and walked out.

The next morning, Brezhnev and Dubcek "agreed" to convene a
meeting of Warsaw Pact leaders in Bratislava on August 3. There, said
Smrkovsky, the Soviet leaders presented a list of demands: that
Frantisek Kriegel (whom one of the Soviet delegates in Cierna had
derogatorily called a "Galician Jew") and KSC Central Committee
Secretary Cestmir Cisar be fired; that the Dubcek regime quash the
effort to reintroduce a Social Democratic party in Czechoslovakia;
that KAN, the Club of Committed Non-Party People, and K-231, the
club of former political prisoners, be banned; and that the mass media
be "streamlined" to impose effective restraints on their output. In
Smrkovsky's view, the invasion ultimately took place because the
Czechoslovak leaders felt unable to accede to all these demands.

Smrkovsky related the following personal account of the Warsaw
Pact invasion and its aftermath. Early on the morning of August 21,
1968, Soviet troops entered the KSC Central Committee building in
Prague and arrested the Czechoslovak leaders who were there, including Dubcek and Smrkovsky. They were flown the same day in two
groups to destinations in Poland and the Soviet Ukraine. Eventually,
they were brought before Brezhnev, Kosygin, and Nikolai Podorgny in
the Central Committee building of the CPSU in Moscow to become
part of the official Czechoslovak delegation. In the meantime, others
in the delegation, including President Ludvik Svoboda, had arrived in
Moscow on their own. Thus, all the major Czechoslovak figures

(including Kriegel, who was originally forcibly detained nearby) were present and participated in the ensuing negotiations with the Soviet leadership in the Kremlin.

The agreement which emerged — an unequal treaty if there ever was one, negotiated and signed under duress — was resisted by some Czechoslovak leaders more than by others. Dubcek alternately argued and collapsed; Cernik argued and cried; Smrkovsky gritted his teeth. The account pictures President Svoboda as the most subservient to the Soviets and haughty to his colleagues, while Frantisek Kriegel was the most resistive to the Kremlin leaders' demands. Svoboda, a general, treated Kriegel as if he were a lowly private; but Kriegel — alone among the Czechoslovak representatives — refused to sign the document of capitulation. In spite of the risk to himself, Kriegel in fact never stopped insisting that the "agreement" be taken to Prague for the approval of the KSC Central Committee and the National Assembly before he would add his signature. According to Smrkovsky, the Soviet answers to this and similar suggestions were blunt and threatening. Boris Ponomarev, Secretary of the CPSU Central Committee, kept saying:

> If you don't sign today, you will sign in a week; if not in a week, then in two weeks; if not in two weeks, then in a month. . . . But you will sign.

Was the agreement valid under international law? Should the Czechoslovak leaders have signed under threat of personal harm? Did they, by signing, in fact "commit treason" against their country? By asking these questions, Smrkovsky implied answers critical of the capitulators, including himself. Virtually on his deathbed, he wanted to leave as clear a statement on the tragic events of 1968 as an eyewitness and participant could make.[7]

Smrkovsky on the Occupation

The second and earlier interview of Smrkovsky in *Giorni — Vie Nuove* was published on September 22, 1971.[8] At the time it was taped, Smrkovsky was already ill with "a serious leg disease" (but reportedly did not yet know it was cancer), had spent several months in the hospital, could walk only on crutches, and was restricted to his home. Nonetheless, the interview was a vigorous presentation of his astute and incisive views of the imprint of the 1968 invasion on Czechoslovak society some three years after the event. The publication of the interview again reflects the firm, negative stand toward the invasion and occupation of Czechoslovakia on the part of Davide Lajolo and many of his colleagues in the Italian Communist Party.

At the start of the interview, Smrkovsky talked about the dilemma confronting himself, his friends, and other purged party members, government officials, intellectuals, and professionals who had resisted accepting the invasion of Czechoslovakia as an act of "fraternal assistance" on the part of concerned allies:

> They [those purged] are told: agree to the occupation of your country or starve. Once you have been sufficiently starved and your children have nothing to eat, you will give in anyway and change your views! That is the "struggle based on principle". . . . Those who have done so [given in] to protect their families have had to violate their consciences, to live in shame, in a state of moral crisis, hating those who humiliated them. Those who have not [given in] are not allowed to work in [their professional] fields. They work as laborers, mostly on roads. . . . Professors are now stokers; former ambassadors are doormen; physicians, truck drivers; journalists, chauffeurs; etc.

Smrkovsky drew a distinction between conditions today and those of the past: he said he did not think in the 1970s tens of thousands of people could be arrested, put in jail, tortured to obtain confessions, and executed on the scale that characterized the 1950s — mainly because world opinion was more influential now than it had been 20 years ago, and because "the power, experience, and level of information in the international Communist movement are different now." Nonetheless, he said, there were still innocent people being arrested, kept in jail without trial for long periods; and finally tried and sentenced to prison. Moreover, he asked, wasn't it horrible enough to witness the expulsion of more than half a million good party members from the ranks of the KSC? Wasn't it horrible enough to see "the corruption of the school system" and "the stifling of any creative activity?"

Turning to the underlying issues, Smrkovsky launched a discussion of the meaning of "dictatorship of the proletariat" and of "socialist democracy." In current practice, he said, dictatorship of the proletariat meant dictatorship, on behalf of the Communist party, by the Presidium (Politburo) and the Secretariat — institutions of "paid officials." If such a system of dictatorship was necessary immediately after the defeat of capitalism, it could not be so justified in an advanced socialist country after some 25 years of existence. "Where," he asked, was "the participation of the people in decision-making, . . . in the formulation of policies, in the execution and control (of such policies)?" How could workers, intellectuals, economists, scientists, etc., support the regime when their activities were prescribed and controlled "by an inept bureaucratic apparatus?" And where could one perceive "*socialist* civil and human rights?"

On the subject of national independence — or Czechoslovakia's lack thereof — Smrkovsky commented that he did not really understand claims of the priority of "class sovereignty" or "internationalism" or other "similar and equally empty phrases, according to which a nation is asked to surrender its inalienable right of self-determination to others" — an unacceptable surrender even if the "others" were "the dearest of allies." How, in any event, could anyone in a country give away that which did not belong to him but rather belonged to all the people?

Since the invasion, he went on, the successors of the Dubcek regime had devoted great efforts to convincing Czechoslovakia's citizens to accept the occupation of their country "as though it had been manna from heaven." But with what result?

> These efforts, which have been carried [to the point of] blind fanaticism and cynicism, have used up all the energy of the party and its leadership and have isolated the party from the people. All of this has deadened people's activity, destroyed the nation's soul, and brought it to a state of lethargy. You may say: life goes on; industry, transport, and agriculture are operating. Of course they are operating. They always function, under any regime, because people's lives depend on them. But the question is how they function. Being out of agreement with present policies, people act accordingly. Those who do agree represent perhaps 10 percent [of the population] at most. The future will show that this is so. All the propagandists' efforts cannot change this; [the attempt] only makes people angry.

Smrkovsky claimed that in 1968 the people of Czechoslovakia had accepted the Communist Party's brand of democratic socialism "with great spontaneity and more overwhelmingly than ever before." Now the population hoped against hope that things would change, that the Soviet Union and its socialist allies would remedy the mistakes they had committed in 1968 and since. In defining the popular mood, he added:

> Our people are like a doctor who, carefully watching a patient, is

prepared to go into action immediately when a crisis occurs. This is a
situation that does not have to be organized by anyone. It exists.

In summation, Smrkovsky suggested that reconciliation of all
divergent views would be the only sensible and responsible solution to
the problems of Czechoslovakia. There should be no victors or van-
quished but only equals. All fraternal Communist parties and indeed
all international progressive forces should have an interest in such an
outcome. The continued occupation of the Czechoslovak state would
constitute "a barrier not only to socialism in Czechoslovakia but to
[the progress of] the entire international socialist movement," he concluded. "Our cause is everyone's cause."

Dubcek Speaks Out

On March 14, 1974, almost two months to the day after the death of
Smrkovsky, *Giorni — Vie Nuove* published a letter addressed to his
widow from Alexander Dubcek. While essentially a message of sympathy and a eulogy to a departed colleague,[9] the letter contained a
number of statements of political import. Dubcek praised Smrkovsky
at some length; he then blamed Gustav Husak for "launching the
accusations" that had led to Smrkovsky's expulsion from the Communist Party in 1970; he defended the actions of his government in
1968–69; he expressed his continued "failure to understand" why
Moscow had chosen a course of invasion on the basis "of manifestly
nonobjective and distorted information" (presumably a reference to
"intelligence" supplied to the Soviets by anti-Dubcek Communists[10]);
and he avowed that the present leadership in Czechoslovakia had lost
"what counts most" — the confidence of the people.

These themes hinted at the smoldering sentiments that were to find
articulation some months later in the ex-leader's remarkable anniversary pronouncement, which was quoted at the start of this paper
and which merits examination in some detail. Like Smrkovsky before
him, Dubcek spoke in his statement both of his personal situation and
of his broader concerns about the course of Czechoslovakia's current
development.

Dubcek, like most of his colleagues, had been expelled from the
KSC and, at the time he wrote, was earning a living as an inspector in
the Bratislava machinery park of the Slovak Ministry of Forests. In the
first part of his message, he complained in bitter terms of sustained
harassment by the secret police. They followed him everywhere, day

and night, he wrote. Why? Because he, like many others who "differ from the present party leadership" merely on the way to achieve "implementation of party policy . . . and [on] the method of solving difficulties" (i.e., on the means not the ends, of Communist rule), were now considered "traitors." "Surveillance webs have been spun, and informants have been planted not only where I work but throughout the society." This harms and frustrates the party and "replaces its main function [of leadership] by rule based on coercion." "This may be a simpler way to govern," he commented, "But it is extremely harmful to socialism [because] it spreads corruption, two-facedness . . . apathy . . . fear . . . suspicion . . . police-informing . . . hypocrisy." In a word, simple use of naked force could not replace open discussion; force should not be the means used "to obtain agreement with party policies and proof of their correctness in the party and society."

Dubcek next turned to Czechoslovak-Soviet relations. These were unfortunately in worse shape than ever before, he concluded. He said that he did not blame the Soviet Union or the other Warsaw Pact members for the 1968 invasion. Instead, he blamed those who allegedly invited the Warsaw Pact forces to intervene in Czechoslovakia:

> If the invasion by Warsaw Pact armies did take place at the invita-
> tion of a *self-appointed group* of Czechoslovak representatives, as was
> announced by the official Soviet news agency, then I can only call such
> a procedure factional activity, to put it mildly, and the consequences
> of this step an affront and humiliation, the moral and ideological
> humiliation of the entire Communist Party of Czechoslovakia and of
> the Czech and Slovak nations. It was (whether [those responsible]
> intended it or not) a heavy blow to the Communist movement.[11]

Dubcek asserted that Czechoslovak-Soviet relations used to be warm and friendly, "with a long tradition behind them . . . strengthened by gratitude for our liberation." Yet, hundreds of thousands of Czechs and Slovaks who had fought side by side with Soviet troops and parti-sans were now labeled "anti-Soviet" and so were more than half a million expelled Communist Party members. "If people of this caliber are so branded, what other result can it have than to evoke anti-Sovietism in the minds of other members of the party and the public?" Nonetheless, Dubcek expressed his conviction that the future of Czechoslovakia could lie only in the community of socialist countries, and that cooperation with the Soviet Union "was and will continue to be the foundation of Czechoslovak foreign policy."

Referring further to the internal effects of the national and party purges of 1969–70 and after, Dubcek said they had caused "immense damage." The creative intelligentsia — those who could contribute most to cultural growth — had been hit the hardest, he reported. At the same time, the workers — "the nominal bearers of power" — were being "thoroughly manipulated," and "workers' participation in public affairs and state administration has been reduced to a minimum, insofar as it exists at all."

Dubcek characterized the present system of rule in Czechoslovakia as "an amalgamation of political, ideological, organizational, personnel, and other instruments and measures wielded primarily from a position of strength." Declaring in a dramatic finale that "this system is not compatible with Marxism-Leninism," he called it, on the contrary, a negation, a denial of socialism: "It is a misuse of power and a violation of socialist principles and party principles. It is a violation of human rights."

The widespread foreign coverage of Dubcek's letter when it surfaced in the spring of 1975 brought about an instant reaction in Czechoslovakia, Eastern Europe, and the USSR.[12] Gustav Husak denounced Dubcek as a counterrevolutionary, not worthy of the "humane" treatment he had received, and asserted that in the wake of Smrkovsky's death, Dubcek was attempting to assert himself as the new leader of a campaign against the Czechoslovak people. Husak offered Dubcek permission to leave the country and go to Sweden (singling out the latter, no doubt, because Swedish Prime Minister Olof Palme had expressed sympathy for the former leader); but Husak made clear that if Dubcek decided to stay in Czechoslovakia, he would have to "conform" and cease to be one of the "open traitors."

Since Dubcek undoubtedly anticipated such a reaction, why did he write his letter? He could hardly have expected to convince the authorities that he was not a betrayer of the socialist cause or to persuade them to free him from police surveillance (after all, he wrote in his letter, even General Svoboda, who was then President of the Republic, "had the habit of placing transistor radios at the windows before opening any discussions. . .") — indeed, he probably knew he was inviting further harassment. In this writer's opinion, Dubcek wrote his letter to offer additional public testimony to the depressing, unstable, unsolved state of affairs in Czechoslovakia. His points are those that have been made before and since in other statements coming out of Czechoslovakia: that Husak's rule has not been successful; that the people have been driven into apathy and passivity but not won over; that the USSR has lost a true friend in Eastern Europe; and that there exists a responsible, articulate and well-informed opposition to the present regime.

The Assault on Talent

A similar sense of a country in distress and in a tight police grip was conveyed by the other aforementioned messages that surfaced in April and May, 1975. The first of these was an open letter to Gustav Husak written by Vaclav Havel, the well-known creator of the "theater of the absurd" in Czechoslovakia — in happier days a prolific, talented playwright but now reported to be a brewery worker in Prague. Havel's letter, copies of which were sent to foreign correspondents in Prague around the third week of April,[13] avowed that the situation in Czechoslovakia was becoming dangerous. "Pervasive fear" was the universal phenomenon — fear of being persecuted or harassed, of losing employment, of being left without any means of support. No one was exempt. The secret police, "omnipresent and omnipotent," was enforcing this system of rule by fear. The manifest presence of the police was in itself "enough to affect one's natural behavior and to induce the habit of pretense." Havel charged that Husak, relying on this system to maintain order, was concerning himself primarily with the party's external image at the price of internal disintegration and societal, spiritual, and moral decay. He concluded with the admonition: "How terrible the hour of truth that may come at some future time!" Like Dubcek, Havel was severely denounced in the Czechoslovak press after his letter was published in the West.[14]

Several specific examples of police intimidation were reported by novelist Ludvik Vaculik in a letter sent to Western news agencies in Prague and published abroad in April, 1975. Vaculik wrote that his home had been subjected to an extensive police search and that some 80 objects — including notebooks, recording tapes, manuscripts, and notes — had been confiscated. According to the novelist, similar searches and confiscations had taken place about the same time in the homes of other prominent individuals, including writer Ivan Klima, historian Vilem Precan, literary critic Ivan Kadlecik, and economist Venek Silhan.[15] Later reports reaching the West suggested that these searches were related to a wave of raids and arrests directed against former high-level members of the Dubcek regime — particularly Zdenek Mlynar, ex-secretary of the KSC Central Committee; and also Jaromir Litera, former head of the Prague party organization; Robert Horak, ex-secretary of the National Assembly; and historian Jan

Kren. Reportedly, the object of this whole series of searches was a document of some 200 pages prepared by Mlynar comparing the political situation in Czechoslovakia before and since the Warsaw Pact invasion, in defense of the Dubcek program and in condemnation of Husak. (Dubcek was said to have sent excerpted copies of Mlynar's document to Erich Honecker, the party chief of East Germany, and Enrico Berlinger, the Secretary General of the Italian Communist Party, requesting that it be made public at the conference of European Communist parties which was then slated to take place in East Berlin in the summer of 1975.)[16]

The effects of police searches and harassment were movingly described in the open letter sent by Marxist philosopher Karel Kosik to his friend Jean-Paul Sartre in May, 1975. In the letter, Kosik accused the Czechoslovak police of violating his "fundamental human rights" and especially his "right to protect his dignity and not to have to give in." He related the hardships that had befallen him: he had been expelled from the KSC, dismissed from his university chair as Professor of Philosophy, deprived of his post of Deputy Director of the Academy of the Academy of Sciences, and even forbidden to drive a taxicab — presumably because he could conduct secret conversations with his passengers. The ongoing intimidation to which he was subjected reached a high point in April, 1975, when the police invaded his apartment, plundered his library, and confiscated a manuscript of some 1,000 pages which he had written. Kosik, now unable to express his ideas even in private, became "simply nothing." Unable to write, unable to support his wife and two children, he was "dead," he wrote to Sartre.

Kosik's case became a *cause célèbre* at the Congress of the International Hegel Society in Stuttgart, Germany, attended by some 900 philosophers and guests from all over the world at the end of May. Jurgen Habermas, the principal spokesman of the philosophy of the Left in West Germany, spoke to the Congress about the case with authority and compassion: If men like Kosik were deprived not only of their livelihood but of the fruit of their private thoughts, what, Habermas asked, was left of socialism?[17] And Sartre, in his reply to the letter, expressed deep sympathy for his friend and for Kosik's country, "invaded and humiliated," a sympathy which "all friends of Czechoslovakia share." "If Karel Kosik is guilty," concluded Sartre, "then all (not only intellectuals but peasants and workers as well) who think the way he does are guilty." In any event, he added, the secret police would not be able to destroy Czechoslovak culture, "try as they might."[18]

The Broader Base of Discontent

The letters and interviews that have emanated from former leaders such as Dubcek and Smrkovsky or from prominent creative intellectuals such as Havel, Kosik, and Vaculik, have attracted a great deal of attention abroad (and have in turn been severely attacked at home). But, as noted above, these headline-making statements represent only a portion of the messages that have been sent out of Czechoslovakia. While considerations of space prevent a comprehensive review, a random sampling may at least indicate the variety, both in form and in content, that has characterized the complaints of Czechoslovakia's unhappy citizens.

A "Ten-Point Manifesto" written by a group of Czech intellectuals, most of them long-term party members, was addressed to the National Assembly (Czechoslovakia's legislature) in protest against the occupation. Two of the writers who hand-delivered the document — former member of the legislature Rudolf Battek and historian Jan Tesar — were arrested in January 1972. The Manifesto itself was published in 1972 in a collection of protest items in the New York journal *New Politics*.[19]

An anonymous "Letter from Prague" was written under the pseudonym George Moldau and published in the British journal *Survey* in the spring of 1973. The letter-writer declared that Czechoslovakia was "a state without leaders" and argued: "We must give a mandate to those 'who are abroad' [e.g., émigrés and exiles] so that they will realize their responsibility and act accordingly."[20]

In a "Letter to Gustav Husak," Milan Hübl, a long-time Communist who identified himself as a former friend of Husak, told a hapless tale of being arrested in late 1971, then released, being "unable to find any job at all" before or after, and finally being rearrested and accused of Zionism on no more grounds than the authorities' presumption that "proof positive of Jewish origin [is] the dieresis on 'a,' 'o,' or 'u' in the spelling of one's name" — while in fact he was, "you will excuse me — 'of pure Aryan descent' . . . ."[21]

A Mrs. Sabatova addressed a letter to foreign Communist parties protesting against the arrest, trial and imprisonment of her husband and three children and asking for intervention on their behalf. The letter was published in December 1973 in the official organ of the British Communist Party.[22]

Among documents issued in 1975, an item of particular interest was an unattributed directory of 144 Czech and Slovak historians, archivists, and curators who had been driven from their professions and in some instances imprisoned. Titled *Acta Persecutionis* (Acts of Persecution), the directory was addressed to the 14th International Congress of Historical Sciences, which took place in San Francisco in August 1975. A brief, unsigned preface stated that the list was incomplete because fear of reprisals had inhibited the collection of information and because persecution was still going on.[23]

In a separate, signed letter addressed to the same congress, one of the listed historians, Vilem Precan (previously cited as a victim of police harassment), described his personal plight. Deprived of his post at the Historical Institute of the Academy of Sciences in 1970, he worked as a receptionist in a Prague restaurant but then was dismissed from that job as well. In April 1975 police raided his home and confiscated part of his library, his files, and his personal diary. Though destitute, Precan said that he would not give in: "I refuse the Czechoslovak state and political leadership the right to treat me as a vassal . . . , to deny me . . . the fundamental right to a decent human and material existence according to my talents, interests, and education. . . . I ask you, esteemed colleagues, . . . to help me as far as you possibly can. I say this quite openly. Without your solidarity and your support, I have no chance of remaining free to resume my work as an historian."[24]

Also in 1975, a Czech playwright named Pavel Kohout addressed an open letter jointly to the German writer Henrich Böll and the American playwright Arthur Miller. He wrote in part: "Czech and Slovak writers have lived and worked in the past seven years under conditions which are a consequence of the year 1968 . . . our actions [in support of the Dubcek program] violated neither laws nor the ethical standards of writers. We have engaged in the defense of ideas which temporarily are being suppressed by the people in power, and that is all. . . . After seven years we and the government have come full circle. We suffer the fate of those who have been silenced. The government suffers the shame of the silencers. . . . In this encounter there are neither victors nor losers, only the defeated."[25]

Other messages, interviews, and documents could be cited. Some have been addressed to Husak or to the National Assembly or both, invoking the writers' constitutional right of petition, while others have been addressed to prominent friends abroad, or to the international Communist or socialist movement, or to the United Nations, or simply to world public opinion. Some have been signed; some have not. Some authors have given permission to publish their messages; others have not. Some have written of personal problems caused by the excesses of the current regime, while others have addressed themselves to the broad problems of the society at large. But all have reflected to some degree a common concern to make known the critical conditions in Czechoslovakia brought about by the invasion and occupation of the country.[26]

The far-ranging content of these messages, in combination with other information, makes it possible to offer a number of general observations about conditions, changes, and developments in that polity. The remainder of this paper will be addressed to several critical questions: the motivations that have prompted the outward flow of protest from citizens, former leaders, and prominent personalities; the indications that such protest reflects widespread disaffection in the population; the frustrated efforts of the Husak regime to "normalize" Czechoslovak society; the impact of international issues and of inter-party pressures within the Communist movement on Czechoslovak affairs; and finally, prospects for the future. The author's observations on these questions are based on his perceptions of the Czechoslovak scene, on a continuous perusal of information emanating from or concerning the country, and on the critical evaluations of other ob-servers inside and outside Czechoslovakia. The analysis below at-tempts to identify the emerging cultural trends, the maturing social conditions, and the evolving political nuances which make up the symbiotic complexities characteristic of Czechoslovakia today.

Dissident Aims and Regime Counteraims

Varied as have been the publicized messages of protest, they fit into a single pattern. They are pleas for help. Realizing that the fate of their country has been determined from outside and that little can be done at home to change that fate, Czechoslovakia's citizens have sent their appeals abroad. This, they believe, is a realistic approach. World public opinion has become an increasingly effective force, and in the late 1970s no country can afford to be insensitive to it. The Soviet Union is deeply concerned to protect its influence in the international Communist movement, which has been split down the middle by the Sino-Soviet conflict. Moscow is equally interested in detente, which might lead to the USSR's acquisition of needed modern technology and to a larger Soviet role on the world economic scene, while at the same time further isolating China. Thus, the stakes are high for the Soviets. Is the occupation of Czechoslovakia worth the censure it costs? The country is already an economic liability to the USSR; should it become a political liability as well? Put the other way, can outmoded strategic and psychological considerations outweigh both liabilities?

The message-writers hope not. They suspect that the occupation itself is not really the main issue; after all, there was no occupation in the 1950s, when the worst party excesses took place. The Warsaw Pact forces will leave Czechoslovakia sooner or later, they believe — among other reasons, because of the sustained pressure on the USSR created by detente and the Sino-Soviet split. But the mere departure of the occupation forces will not mean Czechoslovakia's greater independence. The public will no doubt feel a sense of relief and perhaps may even exact some small concessions from the regime; however, it would be illusory to expect more.

The real target of protest, then, is not the occupation *per se* but the dependence of the present Czechoslovak leadership and *apparat* on the Soviet Union. This dependence is epitomized in the prisoner's dilemma of Gustav Husak, a proud, sardonic, ambitious, lonely, embittered man with an extraordinary passion for power, who has been carrying the burden since 1969. He presents a tragic figure: unable to ignore the proddings of the pro-Soviet ultra-conservatives, but unwilling to give in to them too much; relying for assistance on some whom he earlier condemned, but deserted by many who were former friends; wishing to provide prosperity as a palliative for the new restraints on freedom, but failing to sell the trade-off to the public; willing to permit some relaxation to make things easier, but fearing that if he gives an inch the people will take a mile; wanting to be popular, as Dubcek was, but isolated from the society; needing domestic support, if for no other reason than to increase labor productivity, but, in the process of governing, alienating everyone; unwilling to share power with others, but fearing the concentration of top offices in his own hands (he now wears the hats of Secretary General of the KSC, President of the Republic, Chairman of the Defense Council and of the National Front, and Commander of the Workers' Militia); hoping to improve Czechoslovakia's image abroad, but getting consistently bad publicity instead; and most important, trying to please the Soviet leaders — in the full knowledge that without Soviet support his government would fall apart — but never certain when he may lose their mandate to rule.

In this situation, Husak, his colleagues, and the *apparat* represent a government on the edge of a slippery cliff. As Josef Smrkovsky described their dilemma (in another notable commentary that reached the West[27]): "They think as follows . . . : We cannot do anything against Moscow. . . . We must adjust in fundamentals and must make our country a quiet, *unsuspect* zone, so that we can do something . . . at least partly . . . in the national interest. . . ." The present leaders regard the washing of Czechoslovakia's dirty linen in public abroad as dangerous not only because it centers world attention on their own precarious position but also because it brings Moscow's wrath down upon them. The Soviet leadership would like nothing better than that the world forget Czechoslovakia. Both Moscow and Prague would like the Czechoslovak case to quiet down, to be less visible, less conspicuous — without further reminders of the invasion and occupation.

Thus, the main objective of the Husak leadership has been to put the lid on the pot, to restore "law and order" in Czechoslovak society, a goal requiring the obedience of the masses to the regime at all levels — at the top, in regions, and in districts. As a major means to that end, the leadership has tried its best to create a relatively prosperous society, offering ample food, decent housing, even cars — a society which, satiated (at least by comparison to the past), would not be tempted to get mixed up in politics.[28] However, things have not worked out the way it was hoped. While the economy has in fact registered some temporary gains, the support of the public has not been won over, as many of the protest messages testify. Material comfort, it turns out, is not a substitute for dynamic, principled leadership; and Husak and his cohorts, themselves entirely dependent on Moscow, are hardly charismatic types to whom the masses can relate and respond. Nor can the leaders fall back on ideological appeals: as Smrkovsky put it, Marxist-Leninist beliefs "are to them only phrases which must be mouthed on certain occasions. They know well that the ideology has no social effect [any more], that people simply do not believe in it. It is a game whose rules all know, no one disputes them, and everybody cheats. . . . The few believers in the party are considered either naive, primitive, or stupid."[29]

The Power, the People, and the Party

The power of the leadership is hence based not in the society but in the *apparat*. It is the bureaucratic *apparat* that controls the chain of command at all levels and in all areas and sectors of political and social life — including the party itself, the ministries, the military, the police, the trade unions, the youth organization, and other institutions. Perhaps a quarter-million strong (out of a total population of over 14 million), members of the *apparat,* or *apparatchiks,* are carefully selected by the powerholders to enforce the dictates of the regime. Yet the *apparat* is far from homogeneous. Many *apparatchiks* are simply opportunists who have kowtowed to the new regime and if necessary recanted past "sins" in order to keep a foothold in the power structure. Others got into the *apparat* by sheer chance and are afraid to back out for fear of the consequences. The ultra-conservatives, mostly in top positions, are more pro-Soviet than the Soviets themselves and give Husak a difficult time. Finally, there are the apoliticals who have never trafficked in politics, are without opinions, and thus have survived all the tests of the post-invasion checkups. Willingness to adapt and to obey is their virtue. Generally underlings, they have little ability, experience, knowledge, or skill. All the *apparatchiks* share one common denominator: fear. They support the regime because their own welfare depends on it. They have existentially grown into the regime — in power, they are everything; without it, nothing.

The other social strata, however, appear to be almost universally against the authorities. According to all indications, the workers by a large majority do not regard either the party or the government as theirs; on the contrary, they view the regime as unfriendly. Cooperative farmers appear to be indifferent, concentrating on their own economic and social problems and on their special interests. The economic and technical intelligentsia tend to consider the regime to be without expertise; they work for it because they have no other alternative. The creative and scientific intelligentsia — at least those who made it through the massive checkups and purges — know well for what kind of system they work. Most of them are ashamed that in order to hold onto their positions they lied about their views, while their less compromising colleagues — in many cases including the cream of the crop in particular fields — now dig ditches, if they have jobs at all. The white-collar employees in services such as commerce, trade, health, and transportation do not seem to favor the regime either. As for youth, practically all of them appear to be hostile. So, too, do the many ex-party members. The first and major wave of purges within the KSC was concluded by the fall of 1970, and Gustav Husak announced its results at a December 1970 Central Committee plenum: the party had declined by 473,731 members, or 28 percent; of these, 326,817 members (almost 22 percent) had been expelled, and the rest had quit.[30] Since 1970, additional expulsions have taken place; the total, according to all accounts, is now well over 500,000 people, or almost one-third of the 1968 membership. In many cases, loss of party cards has meant loss of jobs as well. This is a large force. It is composed of people with special political experience, many of whom are highly educated, knowledgeable in the intrigues of the *apparat,* and on friendly terms with some of its remaining members. By all reports, they despise the regime. Could this be the nucleus of an opposition movement in Czechoslovakia? Not at all, according to Smrkovsky: "Undoubtedly, it is a great force which the regime fears. But to speak of it as an opposition is nonsense; it is unorganized, fragmented, watched, and as a whole, powerless." Many of the ex-party members are under surveillance, their telephones are tapped, their mail opened, their visitors photographed, their apartments bugged.[31]

What has all this done to the Communist Party, the "leading progressive force" in Czechoslovakia? According to the message-writers, the KSC is not only an organization of the past — obsolete, regressive, stultified, and doctrinaire — but "a gross and insolent destructive force, an annihilator of human rights and civil liberties." People in an advanced socialist country such as Czechoslovakia have a different image of socialism than that presented by the party today. To them, socialism represents social justice, a modern socialist economy, and a means to achieve national and personal fulfillment. The party is not that. Even in the case of the "proletariat," the party is no longer representative of the workers' struggle for social progress, economic advancement, and other goals of socialism. It is a power organization, pure and simple.[32]

In this country of no real leaders and no real supporters, where everything of importance depends on Moscow but where the rulers keep up the pretense of ruling, where the system keeps running and people keep working and everything more or less keeps going, the citizens seem to feel that they are subject to a kind of historical fate against which any application of will is useless and ineffective. Widespread lethargy, social and political apathy, a mental state devoid of any alternative perspective, pervade the nation. The leadership knows this well. It understands that it has the support of no one except the *apparatchiks.* But it chooses to interpret the national mood of passivity as a political and ideological victory maintaining that the population stands steadfastly behind the new socialist order.[33]

Short- and Long-Term Perspectives

The leadership has attempted to shore up its position by fulfilling at least some societal demands. As pointed out earlier, its effort to raise the standard of living has been not unsuccessful economically but has proven less effective politically. To achieve a compromise with the people, the regime would somehow have to salve their wounded national pride and restore a minimal feeling of national integrity — and that is precisely what the leadership cannot do. This is the dilemma of Husak the prisoner: his hands are tied. A relaxation of controls would be dangerous in a double sense; it could provoke a swift, negative response from the Soviet leaders, or it could start an avalanche at home which he would be too weak to contain. He is hemmed in.

There is the further question whether the regime can continue to provide the economic palliatives upon which it has relied in its quest for viability and stability. To further increase the living standard, promote development, and solve a number of pressing economic problems, the government needs the contribution and cooperation of the working force. Yet it is plagued not only by a shortage of labor but by poor labor discipline — as indicated by widespread absenteeism, failure of workers to produce to capacity, and popularization of the motto, "the worse, the better." All this is hardly conducive to party goals. How can adequate popular support be elicited from a withdrawn and alienated society?

To an observer, it seems obvious that the Czechoslovak political system rests on thin ice. There is little to indicate that post-invasion "normalization" is succeeding; yet at the same time it is clear that the system will not suddenly come apart. Developed political communities, whether socialist or capitalist, cannot do without the participation of their citizens, especially those with education, skill, and expertise, in the socio-economic and political process. But the developed polity must in turn provide an environment conducive to such participation. In Czechoslovakia today, many of those who stand apart are the articulate, the educated, and the expert. Their legitimate demands and needs are simply not being met. Yet, these are people the system cannot do without if it wishes to meet its developmental goals.

Can something be done to make the regime more humane, more responsible, more responsive to people's demands? Smrkovsky had some suggestions:

> We must employ all legal means, from the Constitution to individual administrative orders, to insist on the fulfillment of laws. . . .
> Each citizen can take part in this struggle without great risk. This has nothing to do with anything illegal; this kind of activity is directly based on legality. We must behave with the dignity of citizens, must

not retreat before illegal chicaneries, must protect ourselves and our families, utilize everything which can be of assistance in [terms of] legal norms. . . . To overcome fear is the first victory of a person over a system. I consider the *moral stand of individuals,* their inner strength and resistance to pressure, as most important [factors] in our struggle for the future. . . . It should not be possible to throw people out of work without their defending their rights. . . . One should protest a decision of a local national board (MNV), say, in a housing issue, which contradicts the letter and spirit of a law. People should not tacitly accept a decision forbidding their sons or daughters to study at a high school or a university because their father was expelled from the party. . . . One should fight such decisions directly, against the individuals responsible, in order that they be known, their names remembered, and they not hide behind the anonymity of collective organs in their illegal acts motivated by fear, opportunism, or vengeance. . . . In this way, the regime would begin to realize the limits of its might. . . . As I see it, the main task of opposition [lies] not in organizing illegal activity but in the sustained unmasking of the actual powerholders at all levels, in letting the public know who is who among them, who are the worst, and what acts they have committed.[34]

This counsel would be sound if Czechoslovakia were a nation of saints. But it is not. It is a nation of ordinary people who would face the wrath, "the chicanery," or "the vengeance" of the powerful on all levels. The risks are considerable, Smrkovsky's wish to the contrary notwithstanding. A few may follow his advice — and in fact have — but not many. Thus, as has been the case in the past (in 1918, when Czechoslovakia appeared for the first time on the map; in 1938, at the time of the Munich appeasement; in 1948, when the Communist *coup d'état* took place; and in 1968), any change, to be effective, has to originate outside the country rather than at home. It is in this respect that the Husak regime and the Soviet Union are most vulnerable. The letterwriters are right in directing their appeals abroad.

External Factors of Influence

In pursuing Soviet objectives and dealing with problems across a broad range of foreign-policy concerns — including *inter alia,* detente and the relaxation of tensions, international trade, world opinion, and the conflict with China — Moscow continues to rely on the support of the international Communist movement, or of as many Communist parties as can be mustered to its side. To have several Communist parties disagreeing with the wisdom of the invasion and occupation of

Czechoslovakia — and repeatedly bringing the issue up in public — must, at the least, be disconcerting to the Soviet authorities. The parties in disagreement have been circumspect, but they have seldom left doubt where their sympathies lie. The Spanish CP (in exile) has been the most vocal in its denunciation of the occupation.[35] But far more important, from Moscow's view, has been the reaction of the Italian CP, today perhaps the most influential nonruling party in the Communist movement. Although it has not entered into open argument on the subject with Moscow, the PCI has engaged in many public exchanges with the Husak government. L'Unità, the official PCI newspaper, condemned the Soviet invasion in 1968; deplored the purges and political trials of 1970–72; printed a speech which a representative of the PCI was not permitted to deliver at the Czechoslovak Party Congress of 1970; published a long, moving obituary upon the death of Smrkovsky in 1974; carried extracts from the Dubcek letter to Mrs. Smrkovsky; and so forth. In turn, the Husak-controlled press has printed attacks on the PCI and on individual members such as Davide Lajolo for their stand on Czechoslovakia and the government has twice expelled L'Unità's correspondents from the country (in February 1972 and again in December 1973). Of additional interest, the Italian Socialist Party daily, Avanti!, reported that in 1973, Enrico Berlinguer, the Secretary General of the PCI, was asked to deliver a letter from Smrkovsky to Leonid Brezhnev urging liberalization in Czechoslovakia and to reiterate to Brezhnev the PCI's position with respect to the occupation and its negative effect on detente. According to the report, ". . . the PCI did do its part, without publicity. . . ."[36] Again, when Literaturnaia gazeta attacked Lajolo for the anti-occupation statements in his journal, L'Unità defended him on the ground that he "expressed the [same] position that our party has adopted."[37] (The Italian Communists are, of course, crucially interested in detente, since it reinforces the PCI's independent line and promotes its political influence both at home and in the international Communist movement.)

Other parties have also continued to oppose the occupation, some more openly than others. In France, the Communists have been less vociferous than the Socialists in expressing support for anti-occupation sentiments and forces;[38] nevertheless, the PCF claims that it has not changed its negative view of the invasion. The British and Australian CPs have remained critical of the occupation and have on occasion published letters and messages from Czechoslovak dissidents in their official publications. And so did the CPs of Greece, Belgium,

Holland, Sweden, Japan, etc. Among the parties in power, the Yugoslav regime's sensitivity to the Czechoslovak situation is well-known, though not always visible. The Romanians undoubtedly have similar sentiments, given their own probings for more independence (labeled by one Slovak critic a "policy of walking a tightrope" inspired by "excessive nationalism"[39]).

The Czechoslovak experiment with "socialism with a human face," now perceived as an early "Eurocommunist" variant (a major West European Communist parties' aspiration for autonomy and independence from the USSR — calling into question Soviet leadership of the international Communist movement, as well as at home, embracing electoral, pluralistic, parliamentary means as the only road toward socialism) aroused and solidified major West European Communist parties' opposition to the USSR. In turn, their sustained criticism of Soviet policies in occupied Czechoslovakia (and also in Poland and East Germany) touched a Soviet nerve. Soviet power and prestige in Eastern Europe is at stake, and so is the legitimacy of East European governments.

Eurocommunism now, just as Czechoslovakia in 1968, offers an alternative — and attractive — model to Soviet communism. Because of the emphasis on individual roads to socialism, and in view of the major West European Communist parties' historical ties with East European parties, geographical proximity, close association with Yugoslavia and Romania, and growing influence, it is a model fraught with danger in Eastern Europe. This is where it poses the most serious threat to the USSR, and this is why the Czechoslovak Spring was suppressed so brutally. It was indeed perceived by the Soviet leadership as highly contagious.

When Czechoslovak authorities arrested leading dissidents late in 1976 and early in 1977 in a continuing crackdown on signatories of "The Charter 77," a manifesto for civil rights guaranteed by the Helsinki agreement which was published in West European newspapers, L'Unità wrote that "the virulence . . . leaves no doubt as to the spirit and methods with which the Czechoslovak authorities intend to confront the problems posed by Charter 77," and condemned the Czechoslovak government.[40] Similarly, Rinascita said that "the question of the realization of democratic socialism in Czechoslovakia remains unanswered."[41] A PCE spokesman in Madrid was reported to have called "particularly scandalous . . . this lack of freedom of expression in socialist states."[42]

The indignant condemnations were justified. Fundamental human rights, guaranteed by the Helsinki agreement in the signatory states, had been "regretfully" violated in Czechoslovakia. Tens of thousands of citizens have been prevented from working in their professions. Hundreds of thousands of other citizens live in constant danger of losing their jobs. Countless numbers of people fear that if they manifest their convictions, they or their children could be deprived of the right to education. Freedom of speech is suppressed. Religious freedom does not exist. Other freedoms are disregarded. Human dignity is violated. And yet, "Charter 77" is no organ of political opposition — just a set of hopes, aspirations and beliefs.[43]

The relationship between Eurocommunists and dissidents in socialist states are mutually reinforcing. Eurocommunists monitor events in Eastern Europe, and new trends and developments in West European Communist parties are not lost on the East Europeans and the Russians. The dissidents appeal to Eurocommunists for more support, and the Eurocommunists criticize the socialist states for their excesses. True, censorship is still a potent barrier. But enough filters through to suggest that in Eastern Europe, "critical socialists" not only know what is going on but feel less isolated and deserted. For example, in an open "Letter to the PCI from the Supporters of the Czechoslovak 'New Course,'" the dissident writers in Czechoslovakia praised the PCI's "authentic democracy." "Your position constitutes an important component of the effort to give the cause of socialism in the advanced countries of Europe a new impulse and ensure its progress. *It also provides support for the efforts of all those within the socialist countries who are convinced that the further progress of socialist society is the condition for overcoming the deformations that still exist.*"[44]

In the effort to keep the case of Czechoslovakia alive, the post-1968 top Communist émigrés and exiles have also played an important role. They are the vital link between Czechoslovak society and the international Communist, Socialist, and leftist movements. Unlike the post-1948 émigrés and exiles, who were essentially anti-Communists, this new wave of ex-party members has not only broad support at home but a ready forum outside Czechoslovakia. They claim to speak on behalf of the majority of Czechoslovak citizens and are listened to with attention by the Left abroad. Their major aim is to keep as much attention as possible focused on continuing events in occupied Czechoslovakia. In this they have been fairly effective, especially in Europe. There is some irony in the fact that the exiles, working in cooperation with opponents of the regime at home and with foreign Communist parties, seek an objective that is shared by the Husak government — in that both sides wish the occupation forces would withdraw from Czechoslovakia and that the country could become less dependent on the USSR.

Future Prospects

The future of Czechoslovakia depends on a number of factors, not the least of which are changes in the personnel and the attitudes of the Warsaw Pact countries since 1968. In the Soviet Union, half of the Politburo members who pushed for the invasion are no longer on that body; of the three principal hawks, the Stalinist Arvid Pelshe is retiring, and Brezhnev and Andrei Kirilenko now claim to be detente doves. Hungary's moderate and experienced Janos Kadar has become a friend — and perhaps even advisor — to Gustav Husak. Poland's astute Edward Gierek is certainly not another Gomulka. And East Germany's Erich Honecker is also very different from his Stalinist predecessor, Walter Ulbricht.

A number of other factors may affect the course of events. In Czechoslovakia itself, there is Husak's crucial need to gain a measure of popular support, if for no other reason than to improve the low productivity — and low morale — of the numerically inadequate labor force, in a situation where the obsolete Soviet socioeconomic model must still be followed. In the Soviet Union, leadership decisions are bound to be influenced both by domestic problems (notably, kinks in the economic system, recurring crises in agriculture, and growing tensions among nationalities) and by Moscow's aspirations abroad (including detente with the Western powers; increased trade with Europe, the United States, and Japan; and improved relations with other Communist parties, especially with the Eurocommunists, in view of the cold war with China). In Eastern Europe, there have been many changes since 1968 — in governmental structures and foreign policies as well as in leaderships — and, perhaps more important, the area is rapidly becoming a Soviet economic luxury. In the international Communist movement, among the Socialists, and on the Left in general, sharply dissenting views persist with respect to the occupation — views fed by the continuous stream of dissident messages from Czechoslovakia effectively disseminated by ex-citizens abroad. Finally, there is the factor of world public opinion, to which the Soviet leaders are ever more sensitive.

Given all these considerations, is it so unthinkable that doubts about the wisdom of continuing the military occupation of Czechoslovakia might creep into the minds of the Soviet leaders? And might not these doubts beget other doubts about the wisdom of perpetuating the absolute dependency of Czechoslovakia on the USSR — a dependency which weakens the whole socialist community? Might it not occur to the Soviet leaders that in addition to being an economic, political and psychological burden, Czechoslovakia could become an unreliable military ally — a chink in the armor of the Warsaw Pact?

It is true that the Soviet leaders are not known for their creative ingenuity; all things being equal, they would rather leave well enough alone than stir up a potential hornet's nest. The question is whether they can afford the status quo. After all, the message-writers do not propose the Finlandization of Czechoslovakia. What they plead for is a relationship of limited dependence on the Soviet Union, under conditions that would encourage economic and social development, would look toward Czechoslovakia's partnership in an advanced socialist community, and would permit the pursuit of socialist goals as a means of personal and national fulfillment within that community.

NOTES

* This is a revised and updated edition of a study which originally appeared in *Problems of Communism*, 24, No. 6 (November–December, 1975), pp. 26–42, under the title "Messages from Czechoslovakia."

1. Western coverage included publication of the letter in part and with comments in *The New York Times* (New York), *The Observer* and *The Times* (London), *The Guardian* (Manchester), *Frankfurter Rundschau* and *Frankfurter Allgemeine Zeitung* (Frankfurt), *Die Welt* and *Der Spiegel* (Hamburg), *Le Monde* (Paris), *Arbeiter Zeitung* (Vienna), *Dagens Nyheter* (Stockholm), and many other major papers.

2. Havel's letter was sent to foreign correspondents by registered mail, as reported by Reuters and the German Press Agency (*Deutsche Presse-Agentur* — DPA), which released the news items. See also *The Times* (London), April 23, 1975.

3. Published, *inter alia*, in *Il Populo* (Rome), April 30, 1975.

4. Published in *Der Spiegel*, June 9, 1975.

5. The interview was serialized in *Giorni — Vie Nuove* (Rome), February 20, 1975, and subsequent issues. Extensive excerpts were republished in *The Sunday Times* (London), February 23; *Le Monde*, February 22; and *Der Spiegel*, February 24.

6. Emphasis added. Dubcek's present circumstances will be dealt with later. Smrkovsky's political profile is worth a brief note: having gained early prominence in the party, he was jailed in the 1950s and rehabilitated in the 1960s. In 1969, he was relieved of all his posts. In 1970, he happened to read an article in *Rude pravo* (Prague), the KSC's official daily, which referred to him as a former party member; that, he relates, is how he found out that he had been expelled.

7. It is noteworthy that Smrkovsky, upon returning from Moscow in 1968, had the courage to state in a radio broadcast in Prague on August 29: "Our decision-making was not easy. . . . We were aware that the [Moscow] decision may be regarded, by people and history, as unacceptable and treasonous." See the Czech-language émigré journal *Svedectvi* (Paris), 13, No. 49, 1975, p. 13.

8. An English translation is available in J. Jacobson, ed., "Repression and Resistance in Czechoslovakia," a Special Section of *New Politics* (New York), Winter 1972, pp. 83–90. For the official Czechoslovak rebuttal, see *Rude pravo,* September 25, 1971.

9. In explaining his absence at the funeral, Dubcek wrote that he first learned of Smrkovsky's death "through a Vienna news agency dispatch" and that a telegram giving the date and place of the funeral was delivered too late for him to make the trip from Bratislava to Prague.

10. Those suspected of being informers include four current members of the KSC Presidium: Vasil Bilak, who is also a Central Committee Secretary; Alois Indra, Chairman of the Federal Assembly; Karel Hoffman, Chairman of the trade unions organization; and Antonin Kapek, Secretary of the Prague Municipal Party Committee. Two other suspects are Viliam Salgovic, head of the Slovak Communist Party Control and Auditing Commission; and Pavel Aursperg, head of the Central Committee's International Department.

11. Emphasis in the original.

12. For official Czechoslovak reaction, see *Pravda* (Bratislava), April 17, 18, and 19, 1975; *Mlada fronta* (Prague), April 18; *Lidova democracie* (Prague), April 18. Indignant commentary in other Communist countries appeared, *inter alia,* in *Pravda* (Moscow), *Trybuna ludu* (Warsaw), *Neues Deutschland* (East Berlin), and *Nepszabadsag* (Budapest).

13. See footnote 2.

14. See *Tvorba* (Prague), April 30, 1975.

15. See footnote 3.

16. See *Le Monde,* May 13, 1975; *Der Spiegel,* June 16, 1975. Postponed several times, the conference of European CP's took place in East Berlin in June, 1976. The document was not made public. See Kevin Devlin, "The Challenge of Eurocommunism," *Problems of Communism,* 26, No. 1, 1977, pp. 1–20.

17. *Der Spiegel,* June 9, 1975, p. 123.

18. *Le Monde,* June 30, 1975.

19. Jacobson, *loc. cit.,* pp. 55–59.

20. *Survey* (London), Spring 1973, pp. 245–49.

21. Jacobson, *loc. cit.,* pp. 64–67.

22. *Morning Star* (London), Dec. 5, 1973.

23. The directory carried no date of publication or publisher's name but was translated in both German and English and was presumably printed in West

Germany. For news coverage, see David Schoenbaum, "Writing History Brings Purge of Czech Historians," *The Washington Post,* August 24, 1975, p. E3; and Israel Shenker, "Czech Historians Reported Purged," *The New York Times,* Aug. 22, 1975, p. 30.

24. Excerpted from a five-page letter datelined Prague, July 1975, mimeographed and distributed to the 14th International Congress of Historical Sciences, San Francisco, August, 1975.

25. *The New York Times,* Aug. 29, 1975, p. 27.

26. For still other samples, see a collection of protest messages translated into German and published in Jiri Starek, ed., *Briefe aus der Tschechoslovakei* (Letters from Czechoslovakia), Cologne, Verlag Styria, 1974. See also H. Gordon Skilling, "Czechoslovakia and Helsinki," *Canadian Slavonic Papers,* 18, No. 3, 1976, p. 255.

27. Jan Teren (pseud.), "Night Conversations with Comrade Josef Smrkovsky," *Svedectvi,* 12, No. 47, 1974, p. 413. Emphasis in the original.

28. For a useful internal analysis of economic trends, including a lengthy discussion of the standard of living, see Josef Goldman, "The Czechoslovak Economy in the 'Seventies,'" *Politicka ekonomie* (Prague), No. 1, 1975, pp. 1–17. On the improved wage structure, see Jiri Fremr in *Prace a mzda* (Prague), March 17, 1975. For a comparison of Czechoslovak living standards with those of other East European countries and the USSR, see J. Pitlik, "The Differences Are Being Eliminated," *Hospodarske noviny* (Prague), July 19, 1974.

29. Teren, *loc. cit.,* p. 414.

30. *Rude pravo,* Dec. 15, 1970.

31. Teren, *loc. cit.,* p. 423.

32. See *ibid.,* pp. 420–421.

33. *Ibid.,* p. 416.

34. *Ibid.,* pp. 421–22.

35. It is worth noting that the Spanish Communist Party's theoretical paper, *Realidad,* is published in Rome, apparently with the assistance of the PCI.

36. Luciano Vasconi, "Smrkovsky Is Dead," *Avanti!* (Rome), Jan. 16, 1974. In his letter, Smrkovsky reportedly told Brezhnev how delighted he was with the Soviet policy of detente. He could imagine, he said, that not all Soviet leaders were happy with the policy and likened its opponents to those who had insisted on military intervention in Czechoslovakia in 1968. He then offered his opinion that detente could not succeed as long as Czechoslovakia was occupied. He maintained that while the Husak regime had had some success with the economy, it had failed to solve dire social and political problems that were causing widespread tension and increased anti-Soviet and anti-socialist feeling. Thus, Czechoslovakia was rapidly becoming an unreliable ally of the USSR and a weak member of the socialist community. For this reason, Smrkovsky recommended that negotiations be initiated among the Soviet leaders and representatives of both the Husak and Dubcek regimes without delay.

According to another source (quoted in *Briefe aus der Tschechoslovakei, supra*), Smrkovsky wrote this letter in July 1973. He was warned several times by the secret police not to have the letter published in the West. Apparently, he ignored the warning. For a later account, see *Der Spiegel,* Dec. 9, 1974.

37. *L'Unità,* Dec. 14, 1972.

38. Of interest, in November 1971, the Socialist Party held a two-day colloquium on Czechoslovakia and invited Dubcek, Smrkovsky, and Kriegel to attend. None of them was permitted to go, but Smrkovsky managed to send a letter in which he commented on his current situation and on the sad conditions in Czechoslovakia. See *Le Monde,* Dec. 1, 1972, and *Unir-Débat* (Paris), Dec. 10, 1972.

39. *Smena* (Bratislava), March 10, 1975.

40. Jan. 12, 1977.

41. Jan. 14, 1977.

42. *Frankfurter Allgemeine Zeitung,* Jan. 15, 1977.

43. "'Charter 77': A Czech Dissident Manifesto," Text published in the *New Leader,* 60, No. 3 (1977), pp. 11-14. For Czechoslovak reaction to "Charter 77" and the signatories, see *Tribuna,* No. 7 (February 15, 1977); *Lidova Demokracie,* Feb. 18, 1977; *Rude Prave,* February 18 and 19, 1977.

44. *L'Unità,* June 18, 1976. Emphasis added.

Jiri Valenta

THE USSR AND CZECHOSLOVAKIA'S EXPERIMENT WITH EUROCOMMUNISM: REASSESSMENT AFTER A DECADE

Profound and radical changes, unique in the history of the Communist movement, were taking place in Czechoslovakia in 1968 when the country was suddenly invaded by five Warsaw Pact countries led by the USSR. These changes came to a halt when the invasion interrupted and slowly reversed the peaceful revolution, which was conceived and was being carried out by reform-minded forces in the Czechoslovak Communist Party.

The Prague Spring was a natural outcome of the process of de-Stalinization in a country characterized by a well-developed political system and an essentially Western, democratic tradition deeply rooted in the spirit of its people. This has been well demonstrated in the last study of my mentor and friend, the late Josef Korbel: *Czechoslovakia: The 20th Century* (New York: Columbia University Press, 1977). The Prague Spring was also a product of various reformist trends in the European Communist movement; indeed, Czechoslovak reformists are sometimes looked back upon as forefathers of today's Eurocommunists, who, stressing pluralism and human rights, are trying to forge a model of socialism different from that of the USSR. So similar is the process of democratization that took place in the Czechoslovak Communist Party in 1968 to the trend currently manifest in the Euro-Communist parties of Italy, Spain, and France that Spanish Communist leader Santiago Carrillo, an enthusiastic supporter of Alexander Dubček's reformist program, ventured this striking analogy: "If the term 'Eurocommunism' had been invented in 1968, Dubček would have been a Eurocommunist!"[1]

This paper is an attempt to determine the importance of the Czechoslovak experiment and the subsequent Soviet intervention, seen in perspective ten years later. Three crucial questions will serve as guide-lines: (1) How did the events of the Prague Spring affect the Commu-nist parties of Western Europe? (2) What motivated the Soviet decision to intervene and interrupt the Prague revolution? (3) What have been the consequences of the invasion in terms of Soviet interest?

The Prague Spring and the
Communist Parties of Western Europe

The political crisis in Czechoslovakia, which originally appeared to be only a power struggle, shaped up in the spring of 1968 as a movement toward a more pluralistic concept of socialism, christened by the Czechoslovak reformists "socialism with a human face." In terms of pluralism and human rights, Dubček's reformism was much broader in scope than that of Imre Nagy or Wladyslaw Gomulka in 1956, and even than any phase of Yugoslav reformism. Thus, it is not surprising that the Prague Spring was seen by West European Communist parties, particularly the Italian (PCI), the Spanish (PCE), and the French (PCF), and to some extent by reformist groups in Hungary, Poland, East Germany, and even the USSR, as an example of democratic socialism relevant to the European Communist movement.

Seen from the point of view of conservative elements in the Soviet and East European establishments, however, the reformist orientation of Dubček's leadership created a dangerous political situation in one of the most important countries in Eastern Europe and threatened to spill over into neighboring countries, primarily Poland, East Germany, and the USSR. As the Prague Spring advanced into summer, apprehen-sions increased, as did pressures against the Czechoslovak reformists by the leaderships of these three countries.

Simultaneously, support for Dubček's regime grew more vociferous among reformist elements in the European Communist parties, par-ticularly in Western Europe. Fearing that conservative bureaucracies in the USSR and in Eastern Europe would attempt to crush Prague reformism, the PCI, PCF, and PCE mobilized seventeen West European Communist parties on behalf of the embattled country. During the crisis, this coalition of West European Communist parties applied considerable pressure to discourage intervention. Among other tactics, they threatened to convene a separate conference to condemn the USSR. PCI leader Luigi Longo and PCE leader Carrillo figured prominently in efforts to prevent the invasion. The French Communist leader Waldeck Rochet even served as mediator between the Czechoslovak and Soviet leadership, as did Hungarian leader János Kádár. Although Eurocommunist pressure did not deter the invasion, it caused much confusion and procrastination before the decision was taken, and it moderated Soviet behavior both during and after the intervention.

To be sure, much about Soviet and East European management of the Czechoslovak crisis still remains concealed. The Soviet Politburo ordered the invasion only seventeen days after the seemingly successful negotiations at Čierna and Bratislava. The invasion was largely unanticipated by officials in the East and West, most notably by Dubček, the U.S. government, and the governments of NATO. The former Commander-in-Chief of the U.S. Army in Europe, General James H. Polk, recently noted that the United States and NATO "didn't think it would happen and considered the invasion highly unlikely."[2] (I arrived at the same conclusion myself while discussing this issue with several former high officials and advisers in Dubček's government and in the Johnson administration.) Was the decision to invade made under the cover of a perfect deception, or the inevitable outcome of Soviet "reactionary logic," or what the Soviets perceived as a threat from the West?

Earlier investigations into Soviet management of the Czechoslovak crisis and new evidence explored in my recent study[3] give clues to an alternative explanation of the decision to invade. Soviet willingness to negotiate with Dubček at Čierna and Bratislava was not a ruse calculated to induce a false sense of security while plans for intervention were being perfected. Nor was the Soviet decision based on a uniform set of perceptions of national security, such as fears of a "West German Threat." Such a possibility was not taken seriously by Soviet leaders. After Bratislava it was clear to them that the Dubček government would not "deviate" in its foreign policy orientation, and that NATO intervention in Czechoslovakia was highly unlikely. This is not to say that the argument of the "threat from the West" was not *employed* in the debate preceding the invasion in order to create bureaucratic and public support. Soviet behavior during the crisis brought about a curious mixture of restraint, procrastination, and even tolerance toward diversity, on the one hand, and brutal pressures on the other. The decision was shaped by many factors: the bureaucratic interests and perspectives of senior decisionmakers, manipulated information, East European political instability and pressures, intergovernmental games in Czechoslovakia, signals of U.S. noninvolvement, and finally, shaky compromises between various elements in the Politburo. Eurocommunist counterpressure was not sufficient in and of itself to prevent the invasion.

Motives Behind the Soviet Decision to Intervene

In view of the ambiguous nature of the Čierna and Bratislava agreements and the signs of division within the Soviet Politburo, the decision to invade Czechoslovakia seventeen days after the end of negotiations should not have been so difficult to predict. The ambiguity of the outcome of the Čierna negotiation was indicated by the Soviet decision to withdraw Warsaw Pact troops from Czechoslovak territory but not from the Czechoslovak borders. Thus, intervention was still considered as only one of several possible options open to the Politburo. In fact, the negotiations — while seemingly successful — intensified the ongoing bureaucratic tug of war within the Soviet leadership. In the final debate, the interventionists probably based their arguments on the need for the Soviet Union to be in a position to control for the unpredictability of internal developments in Czechoslovakia after the upcoming Party Congress, and to mitigate the impact of the Congress on the Soviet Union, primarily in the Ukraine in the Soviet west, and on Soviet dissidents and reformers. Equally important was Soviet correction for the impact of the Congress on unstable conditions in the Polish and East German leaderships and on polycentric and autonomous tendencies in Rumania.

Some elements advocating intervention, such as the party leaderships in the Soviet Union's western non-Russian republics, the departments of the Central Committee of the CPSU concerned with ideological supervision and indoctrination, the KGB, and representatives of the Warsaw Pact Command, communicated to the Politburo their dissatisfaction with the sudden moderation of Soviet policy and intensified their efforts to reverse the trend. The bureaucrats responsible for ideological supervision and indoctrination used cryptic language to express their disapproval of the Čierna-Bratislava agreement, which did not call for the reimposition of censorship in Czechoslovakia. The interventionists in the non-Russian republics, particularly the Ukraine, similarly communicated their disapproval of the policy of nonintervention and its stimulation of nationalist sentiments.

In not criticizing Czechoslovakia's dismissal of KGB agents from Prague, in whose behalf a high KGB official intervened unsuccessfully with the Czechoslovak Minister of the Interior, Josef Pavel, the Čierna-Bratislava agreement in effect disallowed resumption of the KGB organizational mission in Czechoslovakia, further angering the KGB leadership. The Warsaw Pact Command considered the Bratislava Declaration ambiguous and was especially displeased with the Politburo's decision to withdraw Warsaw Pact troops from Czechoslovak territory. Since the Warsaw Pact general staff believed that during the Prague Spring Czechoslovakia's defense capabilities were seriously shaken and weakened, it saw a threat to the overall organizational mission of the Pact member countries.

Also, pressures from Eastern Europe had increased. Polish leader Gomulka, fearing that appeasement of the Czechoslovak reformists would intensify factional tension at home and possibly even bring about his overthrow at the November 1968 Congress of the Polish Communist Party, advised the Soviet leadership that he could not guarantee political stability in his country if the Soviet Union did not restore order in Czechoslovakia by use of military force. East German leader Walter Ulbricht signaled to Soviet leaders that the "soft" nature of the Čierna-Bratislava compromise might have unfavorable consequences for East Germany. Czechoslovak antireformists saw the agreement as conducive to their defeat at the forthcoming Party Congress on September 9, 1968. The date of the Congress became an important deadline in the interventionists' policy vis-a-vis Czechoslovakia. They signaled to the Soviet Politburo that the Congress would be a "right-wing" takeover and should be prevented.

The Politburo's final decision and the *timing* of the invasion were evidently based on estimates provided by the KGB and Ambassador S. Chervonenko, on urgent reports from Ulbricht and Gomulka, and on communications from the Czechoslovak antireformist leaders D. Kolder, A. Indra, and V. Bilak (submitted around August 14–15) forecasting a "right-wing" takeover at the Party Congress. All this pressure must have lent credence to the interventionists' argument in the Soviet Politburo that military intervention after the Czechoslovak Party Congress would be much more difficult and costly. The Soviet leaders would then be confronted by a united and legitimatized Czechoslovak leadership, backed by overwhelming popular support. Dealing with the impact of Czechoslovak reformism on the Soviet Union and Eastern Europe, should it be validated at the Congress, would be increasingly awkward.

Meanwhile, the United States, caught up in Vietnam, internal racial disturbances, and presidential politics, was either unable or unwilling to take a stand on behalf of Czechoslovakia. This position was implied in public statements by Secretary of State Dean Rusk in July 1968, by the Johnson administration's continued interest in the SALT negotiations, and by the behavior of U.S. armed forces in West Germany during the summer of 1968. In July, strict orders were given to the U.S. Army in West Germany forbidding all activity, including any increase in air or ground patrols on the Czechoslovak borders, which the Soviets could interpret as supportive of Dubček's regime. The final decision, which reversed the moderate Soviet line adopted at Čierna and Bratislava, may have been influenced by the estimated political and military costs of the invasion and by Soviet assessments of Czechoslovakia's willingness to resist outside interference. Czechoslovakia is hardly a nation of Schweiks, but neither is it given to using force to defend its independence, as the 1938 and 1948 crises had demonstrated. Faced with outside pressures and the threat of intervention, Czechoslovakia's leaders were again unwilling to fight. Dubček in 1968 acted just as Eduard Beneš had thirty years before.

Indeed, Czechoslovakia's unwillingness to resist a military intervention was signaled shortly before the negotiations in Čierna and Bratislava by Dubček's dismissal of General V. Prchlík, head of the Security Department of the Central Committee of the Czechoslovak Party. Prchlík was reported to be the only military man in an important position who had suggested military defense as a possible option for Dubček's government in case of a Soviet invasion. After Prchlík's dismissal and during the negotiations in Čierna, it was not difficult for Soviet leaders to conclude that there would be little military risk in using force in Czechoslovakia.

More importantly, as Zdeněk Mlynář, one of Dubček's colleagues, admitted, the basic fault in 1968 lay in the leader's lack of experience in international affairs and in the absence of a sophisticated foreign policy strategy.[4] In fact, Dubček's entire strategy was based on the mistaken assumption that the Soviets were bluffing. His government did not view seriously the bureaucratic tug of war that ensued in the USSR after Bratislava. (Western observers made the same erroneous calculation.) Had the Czechs conceived a more realistic diplomacy and prepared themselves for possible resistance (as did the Yugoslavs in 1948–1949, the Poles in 1956, and the Rumanians since 1964), the divided Soviet leadership might have perceived a sharp increase in the cost of invasion, thus altering the debate in the Politburo. The Politburo would then have had to choose between limited war against a "socialist ally," with the danger of a possible spillover into West

Germany and a confrontation with NATO forces, and nonintervention, with the problem of East European and domestic containment.

The available evidence suggests that the final decision to intervene was reached over objections from some Soviet decisionmakers. It was arrived at by a majority in the Politburo after the precarious poise between the coalitions had changed in the bureaucratic tug of war. This raises the possibility that a firm posture on the part of the Czechoslovak leaders might have altered the debate in the Kremlin. Unprovoked preparations for resistance might, of course, have triggered an earlier invasion, as some analysts believe. Yet I believe it to be more probable that a firm posture would have increased the costs of invasion sufficiently to discourage the Soviet Union altogether.

Costs and Consequences of the Intervention

After a decade it is difficult to reassess the costs of the Soviet intervention in Czechoslovakia. They have been much greater, it would seem, than a superficial judgment might suggest.

The invasion adversely affected the Soviet posture in world politics. The immediate consequences for Soviet relations with the West were rather unpleasant. The invasion was a temporary setback for the Nonproliferation Treaty, providing arguments for opponents of the Treaty in the U.S. Senate, and delaying approval of the Treaty in several other countries, including West Germany and Japan. More important, it prevented an early beginning of SALT negotiations between the superpowers. President Lyndon Johnson refused to go to Leningrad in October to launch SALT talks with Soviet Premier A. N. Kosygin. President Richard Nixon hesitated for some time to start SALT. Hence, the talks were delayed until the late fall of 1969 (and then entered into at a lower level than the chiefs of governments), more than a year after they might have begun. It may well be true, as was suggested to me by former Secretary of State Rusk, that the momentum in limiting strategic weapons was lost. The delay, in effect, pushed both superpowers into an enormously expensive program of MIRV technology. (Almost simultaneously with cancellation of the Kosygin-Johnson Summit, the tests for perfecting MIRVs began.)

Moreover, the invasion of Czechoslovakia had unfavorable effects upon Soviet policies vis-a-vis Western Europe and Japan. By creating an impression of Soviet unpredictability, it enhanced, at least temporarily, the authority of NATO and its disposition in Europe and the Mediterranean. So Czechoslovakia (a country whose takeover by the

Communists in 1948 provided one of the major reasons for the West to organize NATO) produced in 1968 an element of cohesion and reinvigorated the NATO Alliance. It also contributed to a temporary reversal of the trend in several West European countries, as well as in the United States (under the advocacy of Senator Mansfield), toward a reduction of forces in Western Europe. The intervention also, at least temporarily, mobilized sentiment against the proposed European Security Conference (ESC) and jeopardized the Soviet leadership's hope that the ESC might be convened in Helsinki in the first half of 1970. Equally, it was a temporary blow to Soviet *Westpolitik* in general, and rapprochement with West Germany in particular. Efforts toward cooperation with the West German Social Democrats, led by Willy Brandt, had to be put off. Brandt called the invasion "an earthquake that changed the European scenery, opening fissures and destroying bridges."

In truth, although the invasion called Brandt's *Ostpolitik* and its significance for East-West relations into question, the overall effect was not so traumatic as originally feared. But it slightly altered the military balance in Europe, adding some four or five Soviet divisions to the Warsaw Pact forces and causing several Western countries to increase their military budgets. It also drove some increasingly independent West European countries back closer to the United States. De Gaulle's vision of a grouping of European states led by France, which would provide a counterbalance to both superpowers, was put in jeopardy.

Similarly, in Japan the intervention had adverse consequences for the opponents of extending the U.S.-Japanese Security Pact, which was to expire in 1970. The invasion also had negative effects on Soviet relations with most developing countries in the Third World, which (with the exception of a few Arab countries) unanimously condemned it. Even Egypt, at that time the recipient of Soviet military and economic aid, was ambivalent in her attitude, suggesting, as Radio Cairo put it, that the USSR had been forced to intervene only "because of the part played by world Zionism." In Africa and Latin America the invasion confirmed the fears of many leaders that the USSR was just another imperialist power. Only Mali and Cuba reacted positively to the invasion.

Overall, the Soviet intervention did not significantly alter the general pattern of East-West relationships. The Soviet leadership succeeded in keeping the total political costs of the intervention relatively low, certainly lower than expected. This happened partly because of

Czechoslovakia's inability to actively resist Soviet forces and partly because of the Soviet use of minimal force in implementing the invasion.

The West correctly interpreted the intervention as a purely defensive move, aimed at preserving political stability in the Soviet Union and Eastern Europe. In an era of "thaw" or "cold dawn" and of disillusionment with U.S. conduct in Vietnam, Western mourning over Czechoslovakia did not last long. The West forgot Czechoslovakia with less difficulty than it forgot the intervention in Hungary in 1956. The intervention did not reverse detente, but only contributed to an alteration in its character. Instead of change through rapprochement, detente became rapprochement without change — at least during the terms of President Nixon and President Gerald Ford (from 1969 to 1976). Eventually, East-West negotiations resumed as if the intervention had never taken place. The Soviet posture of deterrence was strengthened and its policy of *Westpolitik* made more credible. In 1971, the Soviet leadership successfully concluded a Soviet-West German treaty; in 1972, the SALT I Agreement. Finally, in 1975, it scored with the convening of the European Security Conference.

The invasion had its most profound effect, not on East-West relations, but on the Communist movement as a whole. The post-invasion climate brought about an increasing willingness on the part of many Communist parties to challenge Soviet authority.

The immediate casualties included the long-awaited Third World Communist Conference, which had been scheduled for November 25, 1968 (aimed mainly at a showdown with China, prior to the Ninth Congress of the Chinese Communist Party). In fact, the invasion seriously weakened the authority of the Soviet Union in its struggle with China. Even though the Third World Communist Conference was finally convened in June, 1969, the Czechoslovak issue remained a serious obstacle to consolidation of Soviet supremacy in the international Communist movement. In any event, the Conference did not become a platform for expelling China from the Communist movement. On the contrary, serious indications appeared at the Conference that Czechoslovakia's aborted revolution and the invasion were encouraging a worldwide trend — excluding perhaps some countries of Eastern Europe — toward greater independence of Communist parties from Moscow.

In Eastern Europe the intervention led to a temporary worsening of Soviet relations with Rumania and Yugoslavia, and provided one of the reasons for the formal withdrawal of Albania from the Warsaw

Pact. After the invasion, the Rumanian and Yugoslav governments developed concepts of total national defense whereby newly created "socialist patriotic guards" in Rumania and a "territorial defense force" in Yugoslavia would fight alongside regular army units in an in depth defense to repel any invaders.

To be sure, the invasion demonstrated to the Communist ruling parties the narrow limits of autonomy set by the Soviet leadership. Stability in Eastern Europe was actually increased by the Soviet invasion, since it served as a warning against repetition of the Prague Spring. But this will probably not be a lasting stability. The invasion decidedly did not signal an end to liberalistic trends in Eastern Europe. Thus, Kádár's regime in Hungary has managed to maintain the economic reform (NEM) initiated in 1968 despite attacks by Soviet and East German economists. Also, Kádár has been able to preserve some domestic flexibility.

In some parties in Western Europe, and in Japan, the invasion has encouraged a tendency toward "socialism with a human face" even exceeding that witnessed in Czechoslovakia. For several important Eurocommunist parties (Italian, Spanish, and French), the invasion, which Ernest Fisher christened a manifestation of *Panzerkommunismus,* served both as a warning about the Soviet version of "proletarian internationalism" and as an inspiration, if not a model, for the road toward pluralistic socialism. As Carrillo pointed out: "The Soviet invasion of Czechoslovakia in 1968 was the last straw." With it, "any idea of internationalism ended for us."[5] Indeed, the Czechoslovak experiment seems to be continuing in several of the Communist parties of Europe. As the leader of the Italian Communist Party, Enrico Berlinguer, recently implied, membership in NATO may offer Western European Communist parties the possibility of developing a brand of communism that the USSR and its East European allies strangled in Czechoslovakia.

The split with China deepened as the Chinese leadership exploited the crisis to condemn the Soviet Union. Chinese leaders appear to have believed, correctly or incorrectly, that the intervention was the manifestation of a new Brezhnev doctrine of "limited sovereignty" seeking to justify Soviet intervention in any socialist country, including China. The intervention convinced the Chinese of the imperialist intentions of the USSR and served as a catalyst in Chinese domestic and foreign policies. The border incidents in 1969 confirmed these convictions. The Chinese leadership reacted to the intervention by curbing the Cultural Revolution, reinforcing the Sino-Soviet borders, and establishing better relations with the United States as a counterbalance to the perceived Soviet threat. The Czechoslovak issue was the first on which the United States and China had agreed in two decades, and it was followed by an improvement in relations.

Finally, and perhaps most importantly, the intervention was costly to future Soviet relations with Czechoslovakia — the only country in Eastern Europe, except for Bulgaria, without a long-standing anti-Russian tradition. As late as 1977, the situation in Czechoslovakia had not become fully stabilized. Indeed, as Charter 77 (a Magna Charta of broad coalitions of Eurocommunist-minded and noncommunist intellectuals, politicians, and workers, including several members of Dubček's leadership) has demonstrated, the incidents of 1968 were not accidental. Czechoslovakia, which is less immune than any other country in Eastern Europe (with the exception of Hungary and Poland) to the ideas of Eurocommunism, still shows no signs of accepting "normalization." It is not only a country with problems of great relevance to Western Europe, but one in which thousands of Dubček's former supporters from "the party of the expelled" see themselves as sharing the ideals and goals of the Eurocommunist movement, to which they look with hope and expectation. Also, the more relaxed East European leaders, such as Kádár in Hungary, view Eurocommunism as a promise and boon to domestic flexibility rather than as a threat.

A statement on the definitive effects of Eurocommunism upon Eastern Europe would be premature at this point. It is possible that Eurocommunism did not die with the Prague Spring in 1968, and that it may reappear in Czechoslovakia or in Poland, the countries most obviously susceptible to its spirit. Certainly, the continuation of the Czechoslovak Eurocommunist experiment in several West European Communist parties, at times even bettering the original, reflects its historical significance. Thus, Czechoslovakia's "socialism with a human face" serves, if not as a model, then perhaps, as "the most specific point of reference for the rejection of the 'Soviet model' of socialism" and for "the rise of the new trend in the Communist movement which has become known as Eurocommunism."[6]

Like Prague reformism, Eurocommunism may become a strong attraction for reformists in various East European countries and in the USSR itself. In fact, the Eurocommunists already go much further than Dubček and his reformists in their program of pluralistic socialism and in their interventions on behalf of reformists and dissidents in Eastern Europe and in the USSR. (Examples are the Eurocommunist outcry in support of the reformists and dissidents in Eastern Europe in 1976 and 1977; and support of the signatories of Charter 77 in Czechoslovakia, of the Polish workers arrested after food riots in June, 1976, and of the Eurocommunist-minded East German dissidents Wolf Biermann and Robert Havemann. Even in the USSR, such reformists as A. Sakharov and R. Medvedev, who saw in the Prague Spring of 1968 a hope for the USSR, harbor the same hope regarding Eurocommunism.) Ironically, Eurocommunism, which emerged as a by-product of the Soviet invasion, presents an even greater menace to the USSR today than did its Czechoslovak variant in 1968. Now as then, a new political specter is haunting the USSR — the specter of Eurocommunism.

NOTES

1. *L'Unita* (Rome), July 14, 1977.
2. General James H. Polk, "Reflections on the Czechoslovakian Invasion, 1968," *Strategic Review*, V, 1 (1977), pp. 36–37.
3. See my article "Soviet Decisionmaking and the Czechoslovak Crisis of 1968," *Studies in Comparative Communism*, VIII, 1 & 2 (Spring/Summer 1975), pp. 147–173. Here I also draw on new research from my forthcoming study of the Soviet invasion of Czechoslovakia.
4. Zdeněk Mlynář, *Československý pokus o reformu 1968* (Cologne: Index-Listy, 1975), pp. 237–238.
5. Santiago Carrillo, in his *Eurocommunism and the State*, quoted in *Time*, July 11, 1977, p. 32.
6. Manuel Azcarate, "Europe and Eurocommunism," *El Pais* (Madrid), July 3, 1977.

APPENDIX

FACTS ABOUT CZECHOSLOVAKIA
compiled by Ben Shomshor

Official title of the political unit: Czechoslovak Socialist Republic
Date political unit created: October, 1918
Date political unit became Communist: February, 1948
Name of political party currently in power: Czechoslovak Communist
Party
Form of government since 1948: Socialist Federal Republic

Common boundaries with other political units: Federal Republic of
Germany, German Democratic Republic, Poland, Hungary, Austria, Union of Socialist Soviet Republics
Area of political unit in square kilometers: 127,869
Seat of government: Prague

Names of chief government and/or political figures:
President, General Secretary of CCP — Gustav Husak
Premier — Lubomir Strougal
Deputy Premiers —
Frantisek Hamouz
Karol Laco
Vaclav Hula
Peter Colotka
Josef Korcak
Jan Gregor
Matej Lucan
Jindrich Zahradnik

Rudolf Rohlicek
Josef Simon
Minister of Foreign Affairs — Bohuslav Chnoupek
Minister of National Defense — Martin Dzur
Minister of Finance — Leopold Ler
Minister of Foreign Trade — Andrej Barcak
Minister of the Interior — Jaromir Obzina

Premier, Czech Socialist Republic — Josef Korcak
Premier, Slovak Socialist Republic — Peter Colotka

Provinces:
Czech Socialist Republic — Bohemia Moravia
Slovak Socialist Republic — Slovakia

Other political parties:
Slovak Communist Party
Czechoslovak Socialist Party
Czechoslovak People's Party
Slovak Revival Party
Slovak Freedom Party

Membership in international organizations:
Warsaw Treaty Organization
Comecon
United Nations (and various UN agencies)

Official languages: Czech, Slovak
Other non-indigenous major languages: Hungarian
Major national minorities: Czech (65%), Slovak (30%), Hungarian,
Polish, Ukrainian, German (5%).

Population: 14,890,000
Population density per square kilometer: 115

Live Birth Rate: 19.8 per 1,000
Infant Mortality Rate: 20.4 per 1,000
Morbidity Rate: 11.7 per 1,000
Leading causes of death:
1 — birth injury, difficult labor and other anoxic and hypoxic condi-
tions (607.9 deaths per 100,000)

2 — other causes of perinatal mortality (581.7 deaths per 100,000)
3 — ischaemic heart disease (270.4 deaths per 100,000)
4 — malignant neoplasms, including neoplasms of lymphatic and
     haemotopoletic tissue (225.7 deaths per 100,000)
5 — cerebrovascular disease (187.6 deaths per 100,000)

Major Religious groups:
Roman Catholic
Protestant
Orthodox

Literacy: virtually 100%

Unit of Currency: *Koruna* (crowns)
Average per capita income in dollars (1974): $2,970
Rate of exchange:
Commercial 5.77 K = $1.00
Tourist 9.80 K = $1.00 (1976)

Gross National Product: $43.6 billion (1974)
National Income: 384.9 billion K (1974)
Imports: $9.1 billion (1975)
Exports: $8.5 billion (1975)

Housing: 4,824,000 dwelling units with average of 3.1 rooms and 1.1
persons per room in 1974.

(all figures below for 1975)
Major Imports:
Animal feeds, cotton fertilizers and minerals, metalliferous ores and
   scrap, petroleum, chemical elements, chemical compounds, agricultural machinery and implements, office machinery, metalworking machinery, road motor vehicles.
Major Exports:
Iron and steel, metalworking machinery, textile and leatherworking
   machinery, electric power generating machinery and switchgear,
   telecommunications apparatus, railway vehicles, road motor
   vehicles, clothing, footwear, agricultural machinery and equipment.

Major Suppliers:
   USSR (32%), East Germany (12%), Poland (10%), West Germany
   (6%), other developed non-Communist countries (25%).
Major Markets:
   USSR (33%), East Germany (12%), Poland (10%), West Germany
   (6%), other developed non-Communist countries (25%).

MANIFESTO OF THE CLUB OF
COMMITTED NON-PARTY MEMBERS*

The approaching 50th anniversary of the Czechoslovak Socialist Republic gives us the opportunity to come out publicly for the ideas that freed our nations and gave the birth of our independent state. With T. G. Masaryk, the founder of our state, we share the belief that states can prosper only as long as they remain faithful to the ideas from which they arose. We profess these ideals in their contemporary, modern form and stress at least three fundamental principles which have now become the backbone of our Club's orientation.

First of all, we think that the bases of all modern political life are human and civil liberties and the equality of all citizens. We regard the defense of these rights against the dehumanizing forces of capitalism, fascism and Stalinism to be the continuation of the unbroken tradition of the struggle of the Czech and Slovak nations for democracy. We see in them the solid base of the Czechoslovak state.

The second point of departure in our political aims is the humanist tradition of culture. Our nation, we believe, found the most significant incentives for their development not on the field of battle, in the struggle for world domination or merely in growth of material affluence, but rather in the sphere of science, art, religion, philosophy, and ethics. In accordance with this international humanist tradition of solidarity, peace and cooperation among men, we find the decisive values not in the nation, race or class, but rather in man's personality and his creative activity.

Finally, our third point of departure, is the impressive idea behind the Czechoslovak experiment — a combination of democratic socialism with far-reaching personal freedoms. The socialist structure of society, the democratic exercise of power and the freedom of the individual are for us the points of departure in our political thought.

---

*May leaflet, distributed by members of KAN, May 1, 1968.

The stormy political developments in our country led to one perhaps paradoxical but certainly logical result: the personal change in leadership was accompanied by other changes. People changed their opinions and attitudes. The whole atmosphere of fear changed to one of trust and good will. The whole structure of political thinking changed as well. Hundreds of thousands of personal revolutions are taking place in the hearts of people. They began to realize that it is unworthy of man to go on trying to sneak through life and not to touch the totalitarian privileges of a small group of people who have seized power. This inner transformation of every thoughtful citizen of our country is for the time being the only guarantee we have that the new development will not reverse itself.

This calls for all like-minded persons to come together in thought and action. It is not only a great surprise but a comfort as well that so many people who are not members of the Communist Party now come out to claim their share of responsibility for the continuing political constructing of the state. Every day the clamor of those who are not Party members grows to have an organization of their own, an organization able to defend their interests. In the abnormal situation of the last twenty years, a hard line was drawn to separate the members of the Communist Party from the others. Those outside were prevented from having a better economic, political and thus social standing. The non-members were manipulated to play the role of the passive, split and least worthy members. Clearly this was a direct violation of basic human rights as formulated in the United Nations Declaration, and a violation of our national traditions of humanism, democracy and socialism.

We cannot yet discern more precisely or in a greater detail the newly emerging structure of our political life. Yet, there are certain common characteristics that bring together all those who are not in the Communist Party. In spite of differences in their political opinions and attitudes, we find a potential common denominator for joint political activity in our Club. The Club has thus far concentrated its political actions on promoting a minimal program, incorporating the demands shared by all non-Party members. We assume that further normalization of our political life will naturally lead toward forming or reforming of political parties according to ideological principles and philosophies. The Club is not at this time aspiring to this role. It does not yet have legal, organisational, material, and personal prerequisites. Yet we regard that now is the time to bring the key issues to the people outside the Party. Now is the time to take the proper steps to ensure that the fundamental rules of a democratic shaping of political life and power in the state will be observed and respected. All outside the Party are interested today in such a program. This regardless of what the present or future parties may incorporate in their programs. This question will be dealt with later, in connection with the elections.

The minimal rules of the democratic process which inspire the political activity of the Club include the defense of fundamental civil rights, especially the equality of those who are not members of the Communist Party. Unless this principle is accepted it will not be possible to correct the defects of our public life in any significant way or to overcome the crisis in our economy. To us the main issue in the coming months is the equality of those who are not Party members with those who are. Next, democratic elections: the only way we are prepared to imagine them is a secret election from separate lists of candidates put forward by several political parties as well as by independents. A new election law must contain at least these elementary provisions for the free participation of the citizens in the formation of the political organs of the state. The law must provide the citizens with genuine alternatives otherwise the democratization will be paralyzed at the very outset.

Finally, we want an alternative platform not only concerning the elections, but also concerning serious political issues of the day. We want to be an independent political power of a new type, representing the interests and opinions of non-Party members. When we form our own political points of view, we do so not in opposition to the Communist Party, but rather in a parallel effort to reach the common goal of both of us — socialism, based on humanism and democracy.

THE DISCIPLINE OF TRUTH

Appeal to the Central Committee of the Communist Party of Czechoslovakia against Svitak's expulsion from the Communist Party. Published in *Student*, 1968.

Prague: June 7, 1964

Dear Comrades,

I became interested in the program of building the socialist society at the time when I was reaching maturity of thought. In the course of the last almost twenty years I participated actively in the practical formation of this society. I consider my participation in the realization of socialism — especially of its humanistic meaning — a considerable portion of my life's goals and, no matter what the results of the Party's proceeding, it cannot change anything on the fact. At the same time, your decision about my membership in the Party has a considerable meaning because it not only decides about a more or less fictitious offense but about the question of whether the intellectuals with a humanistic orientation are, were, or shall be members of the Party only by mistake or because of "immaturity." I believe that my membership in the Party was neither an error nor an act of tolerating an immature, inferior person among grown-ups but a natural expression of identification of a member of the intelligentsia with the program of active humanism. If some of my opinions were given more attention lately it was because I stressed the humanistic meaning of socialism and called attention to some of the interferences of the direction of the cultural life which are not compatible with the ways of a mature socialistic society. Therefore, I do not take any of my opinions back, since it is impossible to retreat from recognized truth. The discipline of recognized truth is the hardest of all disciplines and is superimposed to the discipline of a Party member. It is not tactical to say this truth, however, truth by nature is not tactical. My "case" is not a question of violating Party discipline, it is a conflict of opinions, a conflict of ideological tendencies, a conflict about the meaning of socialism. I believed and still believe in the value of the human being, in man as a being capable of development and I am convinced that the humanistic effort to protect human values is the most important for the understanding of the meaning of socialism. I was convinced in the past and I am now that the socialistic order is created as protection of the working class and, at the same time, of all the people, against the dehumanizing forces of modern technology, against bureaucratic-technocratic apparatus, against the powers of alienation and the cruelty of wars, that it is created as a social order of democratic institutions, free people and developing individuals. I had, and still have confidence in the possibilities of human reason as the only instrument which can indicate the road for human praxis and I refuse to sacrifice on any altar, the most valuable human value — critical thought. I urged, and still urge people to resist the forces which enslave and fool them, to make people recognize the roads towards social and personal freedom. I do not believe in technique and apparatuses because I believe in people and human values. I think, understand, write and act in agreement with the conviction that the meaning of socialism is the growth of human freedom and refuse any revision of this goal of Karl Marx's teaching.

As a philosopher I consider Marx's concept of man, society and history as much more profound than any other, existentialist or phenomenologist and I reject the assumption that existentialists were my models. Unlike Camus, I am convinced that neither history nor human life is an absurd play but that it has a specific meaning since it is precisely in history where human freedom increases in more perfect social formations. I am attempting to confront the concepts of Marx with contemporary science and philosophy; this creates an unusual effect only because of the long philosophical stagnation, dogmatic rigidity and the falsification of philosophical systems — Marxist and non-Marxist, for the last many years. I shall continue in the study of man that I undertook, for as long as I shall be able to and in spite of possible further obstructions, because I am convinced that the understanding of man by means of modern science is the most important task that a philosopher can give himself. A wise person sets for himself the tasks which can be disrupted only by death but not by people or institutions and therefore, my intention to devote the most productive years of my life to the problems of man as a meaning of socialism, cannot be prevented by any decision. This can, of course, either be facilitated or hindered.

The Central Committee will certainly consider whether it is possible in a mature European country with a developed socialistic system and with an excellent humanistic tradition to methodically discredit someone, dismiss him from work, completely silence in print and punish him by the Party for seven unpublished sentences from his private correspondence, misused by the editors. My conflict is a conflict with persons; it is not and never was a conflict with the perspectives of socialism. The differences in opinion in partial questions are not sufficient reason for expulsion from the Party, neither according to existing rules nor according to Leninist concepts about the Communist Party. I am an intellectual and I am not ashamed to admit it. Intellectuals and workers do not feel shame in front of each other but sympathy because they are united by the same basic interests in very much the same way as pseudo-intellectuals and mercenary mobs are united against workers and intellectuals. We intellectuals have been serving the socialistic humanism for the last twenty years with our head, an organ which is not inferior to other organs which can serve the same cause. I am serving with my head which digested the history of philosophy, politics, literature on Marxism, with a head that does not think so very poorly, and which refuses to give up common sense under any circumstances. I am convinced that the Central Committee of the Communist Party of Czechoslovakia needs such a head to use against dogmatists, that it wants it to work for the interests of the labor movement and humanistic socialism, that, however, it needs this head alive, thinking, thinking in its own way. Can I be a member of the Czechoslovak Communist Party with a conviction that the freedom of thought is an inalienable right of man and with the principle that a philosopher serves truth but does not have to subserve the interests of persons? Decide.

With comradely greetings,
Ivan Svitak

ALTERNATIVES IN WINTER*

Achilles' Heel

> Yes, dictatorship! But this dictatorship consists of applying democracy, not eliminating it. This dictatorship must be the work of the class, and not of a little leading minority in the name of the class. . . .
> — Rosa Luxemburg

1. No matter how paradoxical it may seem today, the concept of the dictatorship of the proletariat was originally understood as a system of popular control, or democracy, and not as the elimination of human rights. Therefore, the dictatorship of the proletariat in the orthodox sense has never been the rule of minority over majority or the rule of the elite over the people but has been designed as an assertion of the interests of the majority in a given society vis-a-vis the state. Historically, the theory of a revolutionary vanguard and party elite is Russian by origin and is connected with the idea of a backward proletariat, unable to be in the lead and always in need of the firm rule of a dictator.

2. During the period of Stalinism, this Russian interpretation of Marxism further developed into a doctrine, according to which the replacement of the class by the party, the party by the politburo, and the politburo by the leader became an indisputable dogma. It is obvious by now that Soviet Communists are no longer revolutionaries but only a technocratic power elite which is afraid of freedom. Revolutionaries of yesterday are frightened by one thing: the movement of masses, the activity of people, the action by the working class, and freedom. In fact, they are frightened by Marxism.

3. The moment we accept the assertion according to which the working class does not act on its own, does not express its own will and

---

*Lecture given at Cornell University, Ithaca, published in German, "Der Winter in Prag wird eisig sein," in *Die Presse,* October 12, 1968.

does not understand its own interests, we have to conclude that the only theoretical solution then lies in the idea of the leading role of the Communist Party, the idea of the vanguard — who are at the beginning revolutionary intellectuals, later Stalinist bureaucrats and finally criminals. At this last point, the concept of a socialist revolution is transformed into an idea of extending Communist power by territorial takeovers and of securing the victory of communism by mass repression, concentration camps and military occupation. Together with the degeneration of the revolutionary, later bureaucratic and finally fascist-type elite, the principal institutions of the Stalinist system — the repressive character of which remains unchanged — degenerate as well, although there might be some periods of thaw producing a false impression of progress toward a more reasonable social system.

4. The concept of Czechoslovak socialism, socialist ownership, political democracy and freedom of information, seemed to be able to overcome the gap between the Soviet and Western systems and to find the simple, old definition of socialism: socialism is the growth of individual and political freedom. This concept, as a solution of the existing deadlock, was first adopted by a small group of intellectuals within the Czechoslovak Communist Party. This group of reformists managed to get strong popular support but at the same time antagonized the Soviet power elite which had been moving for quite some time in the very opposite direction. Moreover, in the eyes of the Soviet neo-Stalinists and technocrats, the Czechoslovak power elite committed a mortal sin: *it asked for support from outside the Communist Party* and thus revised the Stalinist interpretation of the party's elitist role.

5. As a matter of fact, the working class adopted at the beginning a very reserved attitude toward the new program, and the creation of workers councils in factories was accepted with indifference. The working class had a measure of security in the official system; it had adequate material conditions, free time and entertainment. A significant change in the workers' thinking started in the period between the summit meetings in Warsaw and Bratislava, when the foundation of the unity of intellectuals and workers was laid. Not only journalists but people at large began *to act as citizens.*

They were defending their moral integrity, their proper identity and overcoming defects in the structure of the industrial consumer society, defects which can no longer be drowned by the increasing production and consumption.

6. The experience from Prague and Paris can be interpreted as a

revolt against alienation, as a result of a clear recognition that the modern society can get rid of its faults and difficulties only by structural changes. All power elites, both in Czechoslovakia and Russia, are vulnerable in any fight in which they cannot take advantage of the strength of the totalitarian system, but are on the contrary challenged by the superior spontaneity, activity and courage of the people. It is here that the elite fails, because it follows its own theory, according to which political activity is a privilege of a limited number of citizens — Communists only — and the population at large is condemned to the status of a second-rate anonymous mob.

7. Any political theory based on the dialectical contradiction of an elite and a mob, on an unchecked rule of leaders, is unsound. Power elites are slaves of their contempt for people to such an extent that they sincerely consider any manifestation of the people's will and *any dissent as rebellion, indeed as counterrevolution.* In their belief that revolution as well as counterrevolution is *produced* by revolutionaries or counterrevolutionaries, the power elites are looking for the nonexistent "producers" of coup d'états everywhere but in the creativity of people.

8. The more perfect the repressive system of a totalitarian dictatorship, the more it is dependent on smooth coordination of all its parts, and the more complex and vulnerable it becomes. The failure of a part may lead to a total failure. Therefore, the better we know the mechanism of a totalitarian dictatorship, the sooner we recognize its vulnerability, the easier it will be for the future democratic left to apply the superiority of a democratic, popular movement against the power of a modern state. Human freedom is the vulnerable heel of any oppressive regime, however strong it may be.

9. It is precisely in the working class and in the industrial plants that there exists an enormous potential of possibilities, which can be utilized if the intellectuals grasp that it is not enough to be brilliant, witty and humorous before television cameras, but that the issue is to formulate political rights for the workers, the right to strike, to free association in trade unions, to the secret election of managers — rights as important and attractive to them as is to us the demand for freedom of the press or of information. The critical task facing the intellectuals is to unite the intelligentsia and the workers.

10. In the light of the failure of the democratization movement, it must be said that attempts made by the European left in Paris and Prague have ended in a common defeat, because the Communist parties as leading political forces are no longer capable of leading the

leftist movement. To lead a radical democratic movement of intellectuals and workers does not mean to flirt with anti-American slogans and play with student provocations. To lead such a movement means to understand theoretically the perspectives of the industrial society and those processes that are transforming the face of the world. A leftist movement can develop only outside the bureaucracies of Communist parties, only outside the traditional revolutionary schemes. Only then can the movement gain the proper experience that may bring victory in the future.

THE ALTERNATIVE OF STRUGGLE

The Spartan warrior Brasidas caught a mouse and while he was holding it in his hand, the mouse bit him. Brasidas said: "Nothing is so small or worthless that it does not deserve freedom, if it is willing to defend itself." He then let go of the mouse.
— From Greek mythology

1. Freedom has to be defended at all costs at the cost of all values, because freedom is the highest value. There was every justification for using military force to defend Czechoslovakia. Yet, at the crucial moment, with a united nation backing its leaders wholeheartedly, not a shot was fired. What good are armies for if they do not defend the state against aggression?

2. To defend the sovereignty of the state is the primary right and duty of every statesman and every state. State leaders who negotiate with the aggressor and are willing to sign a treaty with him enter on the path of compromise of principles which will lead to the replacement of their sovereign state by a protectorate. Any state that does not defend its own sovereignty, must become a protectorate, a non-sovereign state formation, for the reason that the defense of one's own country is the basic feature, right and duty of a sovereign people. Indeed it represents the very essence of sovereignty.

3. Sovereignty is the basic value of national life. If it is given up, the nation is reduced to a mere geographic and demographic unit. The possibility for an independent economic, social and cultural development is a basic right of all nations, generally recognized and respected. Only fascist-type political parties and states do not recognize such a right and violate it with impunity — at least for a certain length of time.

4. State sovereignty and national independence have been the basic values for centuries of European life. In 1968, the sovereign people of Czechoslovakia did not endeavor after national interests only but also after a humanistic, European concept of socialism. Czechoslovakia

could and had to fight for these basic values of European culture, for its national independence and state sovereignty as these values cannot be given up and surrendered to anyone, as long as the principle of the people as the only source of power remains valid.

5. The military arguments in favor of defending Czechoslovakia are more complex than any legal arguments, but they are equally strong if one stops being obsessed by the argument of the geographic indefensibility of the country. In certain cases, it is necessary to accept the fight even under the most unfavorable circumstances, since the fight for freedom represents a value in itself, regardless of the outcome of the fight. This is the major argument in favor of armed resistance well understood by a small country like Finland in 1940 and by the Hussites five centuries ago, or again by David facing Goliath. In a situation, when all basic rights are threatened, a nation or an individual must fight even without any prospect of victory.

6. Another reason for attempting the defense of Czechoslovakia lies in the nature of military operations, the course of which can never be fully foreseen and cannot bar unexpected developments. A military resistance by the Czechoslovak army might have precipitated a crisis in Moscow, which consequently could have politically cancelled the whole Russian military action. The Czechoslovak leaders committed a fatal mistake by not preparing the country for war and not risking a complete break with the Soviet Union. Faced with the possibility of an open war in Central Europe, the aggressors may not have dared to intervene. Unfortunately, the Russian generals never had to face such a possibility and at all times they were free to decide with impunity whether or not to occupy a defenseless country.

7. Still another reason for deciding in favor of defending Czechoslovakia militarily should have been the historical experience that any appeasement of any aggressor is essentially a political mistake. There is no other way to counter military aggression except by military means. Soviet fascism has to be fought the same way as any other kind of fascism. The occupation of Czechoslovakia is just a prologue and any policy of appeasement has so far proved a worse mistake than any policy of armed resistance. Hasn't the Third World War actually started by the Soviet occupation of Czechoslovakia?

8. To prepare the nation for armed resistance was, in summer 1968, legally justified, militarily possible and morally imperative. The moral consequences of not doing so will prove as harmful to the nation as the occupation itself. No nation can expect to thrive if its backbone is broken every twenty years. Following the political and military failure

will come a moral and intellectual crisis when the reality of what has happened finally sinks in. Once this reality is accepted, a "realistic" policy will lead toward a camouflaged collaboration with the Soviet occupiers and finally toward a complete betrayal of the nation. These are the cruel laws of history.

9. If leading politicians discredit their own program by extensive compromises with the occupation forces, they will undermine the basic national value — faith in the just cause of democratic socialism. At the same time, they will foster a sentimental, small-power complex of affront and humiliation, which will ruin the very principles of humanistic democracy and genuine socialism that helped them take over power. They will also ruin the basic belief in man, justice and commitment to human rights. These consequences will be more disastrous than any war could have been.

10. In defending the idea of democratic socialism, we would not be fighting for any utopia but for Europe's future and its values and for the basic values of man as a free human being. We have never been so close to the great ideas of most prominent men in history as in the summer of 1968. Never before have we been so close to Europe and Marx and never before have we been a more active historical force. At a crucial moment, however, we gave up our fight and forgot the very essence of our history as the history of a European country fighting for its freedom to avoid extinction.

ALTERNATIVE OF COLLABORATION

"If you ask me whether our sovereignty is endangered, I am telling you openly — it is not."
— Dubcek's statement made two weeks before the Soviet occupation of Czechoslovakia.

1. The passive resistance of the Czechoslovak people against the overwhelming force of Soviet tanks in August 1968 was as justified as the earlier decision of the leadership not to defend the country militarily was unjustified. The leaders' unrealistic expectation that the country would not be attacked and state sovereignty not threatened launched passive resistance on a massive scale as the only possible answer to the occupying forces and as a substitute for armed struggle. The great political capital amassed by passive resistance was quickly dwindled by a leadership that accepted compromise and that was not able to make political use of the national unity against the occupation forces.

2. While the passive resistance of the people prevented Russia from establishing a Soviet puppet state, Communist leaders of Czechoslovakia adopted a suicidal policy of compromise and collaboration, thus increasingly liquidating the spirit of the people's resistance. It was not the conservative elements or the Novotnyites *but the reformists themselves who offered a service to the occupation forces by accepting their diktat*. In fact, no need arose for the occupiers to interfere in the internal affairs of the country since they were able to force the Czechoslovak government to execute Soviet policy under the slogan of "normalization."

3. Already during the first crucial week of the occupation the Czechoslovak leaders were confronted with the alternative to fight or to collaborate. They tried to solve the unsolvable dilemma by a series of compromises which legitimized the occupation, limited the sovereignty of the state and the basic human rights. Unlike their leaders,

the Czechoslovak people adopted from the very beginning of the occupation the policy of passive resistance. In view of persisting illusions about national unity and national heroes and because of the personal prestige enjoyed by certain Czechoslovak leaders, it has been difficult so far to realize that passive resistance and national unity have ceased to exist.

4. What a bizarre historical paradox to have the same men execute Soviet orders who had been looked upon as defenders of freedom. Mr. Čestmír Císař, for example, on September 18, 1968 characterized the agreement signed in Moscow on August 26 as "historically logical, politically realistic, sound and mutually useful." Zdenek Mlynar declared on television on September 15, 1968 that the basis for "normalization" had been actually created during the critical night of August 21!

5. The final blow to passive resistance was given by the leadership, the government and the National Assembly. On October 18, 1968, the parliament approved a treaty granting the Soviet army the right to be stationed indefinitely on Czechoslovak territory. The treaty provides for no limitation of the movements of Soviet troops, does not provide for any compensation for damages caused by the occupation forces and gives the occupation itself an ex post facto legal status. This treaty confirms the occupation as something permanent, and thus accomplishes the ultimate objective of the Soviets by an act signed by Czechoslovak "reformists." This is the actual result of a policy based on compromises and represents a new Munich, invalid from the start since it has been imposed by threat of violence.

6. The third and most decisive step was taken a month later at the November session of the Central Committee of the Communist Party which resulted in the creation of a pro-Soviet faction within the Party and the isolation of hesitating reformists around Dubček in order to secure the leading role of the Party bureaucracy in the name of the "new realities."

7. The reformists' choice of collaboration has been a wrong one for at least three reasons. First of all, passivity as such is no program; it can last for a limited time only and it disintegrates quickly, particularly under revolutionary circumstances which call action of any type rather than retreat. Defensive tactics lead to the death of any uprising.

8. Secondly, the alternative of collaboration was connected with the concept of legality. Czech politicians presented theoretical arguments which, however, were consistently refuted by the Soviet actions. Thus, it so happened that the National Assembly turned down the idea of

neutrality as something not in accordance with *legality*, while witnessing the violation by the occupiers of every concept of international law by their mere presence in Czechoslovakia. The Czechoslovak leaders failed to draw the natural conclusion, namely, that a "gangster" can hardly be impressed by legality.

9. Passive resistance, combined with a policy of compromises and the acceptance of Moscow's *diktat,* caused a gradual retreat of Czechoslovak leaders and resulted in Moscow's salami tactics which began to break down resistance and enabled the replacement of reformists by so-called realists and collaborators. Collaboration can neutralize the people's indignation only temporarily. In the long run, however, it will confirm the conclusion that a nation will not be free until it defends its freedom in armed struggle.

10. To be radical means to have faith in reason and not in any new ideology. To be radical means to think truthfully and not to get deceived by illusions. To be radical means to see reality in its truth — and to refute it. If we are revolting against the given reality, it is not because we would not understand the limitations and the "realism" of policy-makers. On the contrary, it is just because we understand them very well. *To resist the existing reality is the only honest attitude which is left after the loss of the alternative of struggle and after the failure of compromises. The alternative of resistance is thus the only alternative left.*

Index

Z